Wave Momentum and Quasi-Particles in Physical Acoustics

WORLD SCIENTIFIC SERIES ON NONLINEAR SCIENCE

Editor: Leon O. Chua
University of California, Berkeley

*To view the complete list of the published volumes in the series, please visit:
http://www.worldscientific.com/series/wssnsa

WORLD SCIENTIFIC SERIES ON
NONLINEAR SCIENCE

Series A   Vol. 88

Series Editor: Leon O. Chua

Wave Momentum and Quasi-Particles in Physical Acoustics

Gérard A Maugin
Martine Rousseau
Université Pierre et Marie Curie, Paris, France

W World Scientific

NEW JERSEY · LONDON · SINGAPORE · BEIJING · SHANGHAI · HONG KONG · TAIPEI · CHENNAI

Published by

World Scientific Publishing Co. Pte. Ltd.

5 Toh Tuck Link, Singapore 596224

USA office: 27 Warren Street, Suite 401-402, Hackensack, NJ 07601

UK office: 57 Shelton Street, Covent Garden, London WC2H 9HE

Library of Congress Cataloging-in-Publication Data

Maugin, G. A. (Gérard A.), 1944–   author.

  Wave momentum and quasi-particles in physical acoustics / Gérard A Maugin, Martine Rousseau (Université Pierre et Marie Curie, Paris, France).

    pages cm -- (World Scientific series on nonlinear science. Series A ; v. 88)

    Includes bibliographical references and index.

    ISBN 978-9814663786 (hardcover : alk. paper) -- ISBN 9814663786 (hardcover : alk. paper)

    1. Wave mechanics--Data processing. 2. Quasiparticles (Physics). 3. Momentum wave function.

4. Elastic wave propagation. 5. Nonlinear acoustics. I. Rousseau, Martine, author. II. Title.

III. Series: World Scientific series on nonlinear science. Series A, Monographs and treatises ; v. 88.

  QC174.2.M38 2015

  534--dc23

                                                                          2015006577

British Library Cataloguing-in-Publication Data

A catalogue record for this book is available from the British Library.

Preface

After a long period of competition, the wave-like and particle-like visions of some dynamical theories seem to have reached an agreement in their useful complementarity. Both serve to describe propagating information via their well founded duality. The wave modelling favours a description of the propagation of information in terms of wave number and frequency. As to the particle model, it pertains to a diffusion of information through certain interactions in terms of momentum and energy. Classically, the duality between the two models is built quantum mechanically. Regarding this competition the interested reader will find it rewarding to peruse the book of B. R. Wheaton[1]. In the present book, we are interested in elastic vibrations in a deformable solid; the relevant particles then deserve the christening of "quasi-particles", the most popular ones being the *phonons*. In a crystalline lattice outside the absolute zero, random motion takes place that corresponds to heat. In a crystalline medium subjected to boundary conditions, the phonon is associated with a *modal* vibration characterized by its frequency. The "particles" just conceived, e.g., phonons, are objects introduced to easily model interactions at this micro-level. But Nature also offers a more global scale of wave phenomena involving interactions that can essentially be understood in terms of a particle or "quasi-particle" description. In recent times, such waves are called solitary waves. These, briefly described, have the shape of a unique strongly localized "wave" of unusually large amplitude moving over long distances at the surface of a fluid or a deformable solid[2]. The allied quasi-particle interpretation is then

---

[1] B. R. Wheaton (1983). The tiger and the shark: Empirical roots of wave-particle dualism, Cambridge University Press, UK.

[2] The interested reader may consult the short historical perspective given by one of the authors: G. A. Maugin (2011), Solitons in elastic solids (1938–2010), Mechanics Research Communications, 38, pp. 341–349.

particularly well adapted and rapidly gives rise to the notion of *soliton*. This notion is a materialisation of the wave that carries the information.

In the present work, we study the quasi-particles that are dual — in the sense of the above emphasized duality — *of elastic waves* that are *known* solutions of elastic wave problems, i.e., in physical acoustics. The interest in this study stems from the potentially associated simple interpretation of the interaction between fellow waves or of the interaction of such a wave with material objects (discontinuity surfaces, defects, inclusions,...) The proposed original approach consists, once we know a macroscopic wave solution, in exploiting the conservation equations of canonical (or wave) momentum and energy, as recently revisited by us in continuum mechanics. This methodology is easily understood. Standard equations (here field equations of elasticity and associated boundary conditions) are used to obtain the dynamical solution (e.g., the celebrated Rayleigh wave for surface propagation). Then another set of continuum equations (so-called conservation laws in the sense of *Noether's theorem* of field theory) is used — such as in a post-processing — to build some other quantities, here those that will principally appear in the conservation of so-called wave-momentum. It is that equation, and possibly in parallel the energy equation, which is integrated over a volume element that is representative of the considered wave motion (e.g., a vertical band of width equal to one wavelength in the sagittal plane for surface waves). This integration will provide the looked for equations (momentum and energy) of the associated quasi-particle, often exactly in the form of a Newtonian "point mechanics", but with a "mass" that may depend on the velocity. This strategy is the one that pervades the whole of "configurational mechanics" in modern continuum thermomechanics[3]— where it is applied to the evaluation of driving forces acting on field singularities, material defects, inclusions, phase-transition fronts, *etc.*

The first chapter should be viewed as a prolegomenon introducing the inclusive notions of *wave momentum* and *radiative stresses* in one dimension of space in the line of Léon Brillouin. This provides an occasion to pay a tribute to this brilliant, sometimes underrated, physicist[4]. The notions of

---

[3]See the book by one of the authors: G. A. Maugin (2010). Configurational Mechanics, (CRC-Chapman & Hall, Boca Raton, FL).

[4]Léon Brillouin (1889–1969) is a French-American physicist of the Nobel-prize calibre whose name is attached to many remarkable advances (quantum theory of solids, Brillouin scattering, Brillouin zones, WKB method, Brillouin–Wigner formula, notion of neg-entropy). He is the physicist who thoroughly clarified the notion of group velocity in dispersive systems (*Cf.* his book with A. Sommerfeld, Wave Propagation and Group Velocity, New York, Academic Press, 1960). He is also the author of a remarkable pio-

Eulerian and Lagrangian descriptions are introduced for this purpose. The contents of this chapter owe much to our fruitful collaboration with the late Alexander Potapov from Nizhny–Novgorod, a famous school of nonlinear physics. The bases of continuum thermomechanics are dealt with in greater detail and more precise mathematical format in three-dimensions of space in Chapter 2 with the necessary reminder on finite-strain and small-strain elasticity theories. Following along the same line, Chapter 3 introduces the general notions of *pseudo-momentum* and *Eshelby stress* — i.e., the time-like component and the spatial part of an energy-momentum tensor — by application of Noether's theorem of field theory when a variational formulation is at disposal (e.g., in pure elasticity). However, a procedure mimicking Noether's identity is also proposed in the presence of dissipation (such as viscosity). Chapter 4 deals with the notion of *action* (energy multiplied by time) — seldom considered locally in continuum mechanics — and elements of the *wave mechanics* expanded by Lighthill, Whitham and Hayes, and recently revisited by one of the authors. This exploits perturbation schemes.

True applications of the notion of quasi-particles associated with elastic wave modes start in Chapter 5 with the problem of transmission-reflection at material interfaces, whether perfect or imperfect. Chapter 6 follows with the application to so-called *dynamic materials*. The latter have the essential property of presenting heterogeneities in both space and time. That is, in addition to the presence of material discontinuities reflecting abrupt changes in inertial and elastic properties, these materials are also subjected to more or less abrupt changes in time such as due to the action of a practically sudden import of energy through an external action. Such systems are thermodynamically open. The phenomenon of interest is very much like the pumping of energy in lasers. The point of view of quasi-particles sheds an interesting light on this phenomenon where the wave-like picture is somewhat artificial. Chapters 7 and 8 deal with a subject of importance in seismology and electromechanical devices: *surface waves* of both elastic and electro-elastic nature. Special emphasis is placed in Chapter 7 on the cases of *Rayleigh waves* (even when perturbed by surface energy or an external fluid), Love waves (multimode dispersive waves), and so-called *Murdoch surface waves* which correspond to the propagation of

---

neering book on Tensors in mechanics and elasticity (First French edition, Paris, 1938; Dover reprint, New York, 1946). A detailed biography and list of publications were given by L. Hilleth Thomas, "Léon Nicolas Brillouin 1889–1969", Biographical Memoir, National Academy of Sciences of the USA, pp. 68–89, Washington, DC, 1985.

a mono-mode shear surface wave. The latter mode can be viewed as a limit of one of the Love surface acoustic modes when the superimposed layer has a thickness that tends to zero while keeping its essential kinetic properties. This limit property is also kept for the view of the associated quasi-particle properties. The inclusion of Chapter 8 may be due to an immoderate taste of the authors for electro-magneto-mechanical couplings. But this chapter, by selecting a rather simple case of coupling (so-called *Bleustein–Gulyaev waves*), provides an occasion to illustrate the possible influence of a nonlinearity in the elasticity of the substrate, and also of a possible viscosity of this substrate. This last case is of particular interest because it yields a non-inertial motion of the associated quasi-particle with a driving force interpreted as friction. Globally, these two chapters introduce the notion of quasi-particles guided by a bounding surface.

Chapter 9 offers a different perspective by unfolding what happens to the notions of wave-momentum and associated quasi-particles in what are now referred to as *generalized continua*. These include three different types of generalization of standard (Cauchy–Navier–Green) elasticity: a weak non-locality accounting for higher order gradients of the elastic displacement, an internal microstructure that accounts for additional degrees of freedom of rotation (such as in Cosserat continua), and strong nonlocality represented by space-functional elasticity constitutive equations. This last point in fact brings us back to the discrete vision of crystalline structures. Finally, Chapter 10, although not intended as an additional treatise on solitons, presents a few original cases which then include simultaneously both dispersion and nonlinearity, by generalizing some of the cases already exposed in Chapters 7 and 8. The corresponding results have often been obtained in close co-operation with other researchers in the period 1985-2002, especially Joël Pouget and Hichem Hadouaj (in Paris), Boris Malomed (now in Israel), and the late Christo I. Christov (Bulgaria, and Lafayette, Louisiana, USA).

In all, we expect that the contents of this work, quite original and somewhat thought-provoking, should be of interest to curious physicists, applied mathematicians, and mechanicians, as we think that they broaden insufficiently studied avenues in the physics and mathematics of wave propagation in deformable bodies.

The final presentation of the text and the beautiful figures are due to the expertise of Ms Janine INDEAU, whom we thank from the bottom of our heart for her kind and efficient co-operation.

Gérard A. Maugin and Martine Rousseau

Contents

Chapter 1

Prolegomena: wave momentum and radiative stresses in 1D in the line of Brillouin

Object of the chapter

This introductory chapter, a kind of tribute to Léon Brillouin (1889–1969) and a prerequisite for the sequel, aims at introducing some basic notions such as those of wave momentum and radiation stresses in an illustrative one-dimensional setting in space. In this view the notions of Eulerian and Lagrangian kinematic descriptions are also introduced in both of which a method of successive approximations (first and second-order approximations) in the neighbourhood of a rest unstrained (and unstressed) state is applied. This allows one to pinpoint the effects of nonlinear elasticity on the notion of wave momentum and radiative stresses in both descriptions. The case of momentum and radiative stress in a thin rod serves as an example, yielding eventually the acoustic levitation effect. This paves the way for the other chapters.

Note. The contents of this chapter are essentially based on the following previous publications: Brenig (1955), Maugin and Trimarco (1991– incorporated in Maugin, 1993, pp. 34–38), Potapov and Maugin (2001a,b) and above all Potapov *et al.* (2005).

1.1 Introduction

In his magisterial doctoral thesis (Brillouin, 1921) which constitutes a foundational work in a then new science — solid state physics —, the French/American physicist Léon Brillouin (1889–1969), emphasizes the required focus on elastic wave propagation to approach the general problem of vibrations in deformable solids. On this occasion, he deals with one of his recurring subjects of interest, the notion of *radiation* (or radiative) *stresses* after previous works by, among others, Lord Rayleigh (1905) (see also Brillouin (1925)). The present introductory chapter deals with this notion in modern terms, but in one dimension of space, together with the allied notion of *wave momentum*, a much debated notion.

In a nutshell, in a one-dimensional setting, the traditional *linear momentum* reads

$$p = \rho\, u_t \tag{1.1}$$

(where subscript t denotes the partial derivative with respect to time and u is the only surviving component of displacement; a subscript x will denotes the partial derivative with respect to the x space coordinate) and describes *mass transport* through the equation of conservation of mass

$$\rho_t + p_x = 0. \tag{1.2}$$

The flux associated with p in the local balance of linear momentum is the standard (Cauchy) stress σ in the equation

$$p_t - \sigma_x = 0 \tag{1.3}$$

in the absence of body forces. For a harmonic motion p is of *first order* in the amplitude U of the motion. But when wave processes are investigated in elastic bodies there are additional quantities that are constant in time, i.e., are *conserved*; these include quantities that are *quadratic* in the wave amplitude. Such are energy, wave momentum and the associated radiation stresses.

The last two notions have sometimes been used in the analysis of wave processes in continua and are considered in works on acoustics in fluids (see, e.g., Westervelt (1957), Zarembo and Krasil'nikov (1966); Danilov and Mironov (2000)). Here the origin of radiative stresses is to be found in the change of wave momentum caused by scattering on inhomogeneities, reflection on obstacles, absorption, or radiation. The existence of radiative stresses is related to the change of the time averaged momentum of the medium transported by the wave. In problems of hydrodynamics and gas dynamics nonlinear equations of motion need to be analysed in order to explain the appearance of the notion of wave momentum. This is attributed to the secondary flows generated by the acoustic field, and one generally has to go beyond the scope of one-dimensional problems and consider sound beams and secondary flows produced by the acoustic field. This is illustrated by the works of Stepanyants and Fabrikant [1989, 1999].

Back to 1D elasticity, the expression of the wave momentum — as we shall see later on — is given by

$$p^w = -\rho\, u_t\, u_x = -p\, u_x = -p\, e, \tag{1.4}$$

where $e = u_x$ is the infinitesimal strain. Contrary to the standard linear momentum (1.1) — that will be later on referred to as the *physical* momentum — and may differ from zero in the absence of deformation wave, we see that p^w vanishes in the absence of strain. It obviously is second-order in the amplitude for harmonic motions. A more precise definition of both wave momentum and radiative stresses will be given in a further section by considering a method of successive approximations in the neighbourhood of a stationary state. Note that since nonlinear terms are necessarily involved special attention must be paid to which representation — Eulerian or Lagrangian — is used for the description of the motion. In a general theory of deformable solids, it is the referential (material) description that is well adapted for the present considerations.

Remark 1.1. The notions of radiative stress and wave momentum are not proper to the acoustics of solids. They show up in wave systems of arbitrary nature, whether electromagnetic (*cf.* Ginzburg and Ugarov (1976), Nelson (1990)) and optical (e.g., Jones and Leslie (1978), Peierls (1985)), in surface waves on water (LeBlond and Mysak (1978); Stepanyants and Fabrikant (1999)), and for elastic surface waves (as largely documented in this book). For electromagnetic waves in a vacuum the wave momentum is one of the integrals of motion for the set of linear Maxwell equations. The situation is much more involved for electromagnetic waves in a material medium, even when the motion of the medium is neglected. The definition of momentum is ambiguous even in this simplified case. The two most frequently used candidates for electromagnetic momentum are those proposed by Minkowski and Abraham early in the twentieth century. Which of these two, or still another one, is more physical was a much debated question (see Eringen and Maugin (1990b, Chap. 3)). Attempts at a sensible definite answer to this question have been proposed by, among others, Nelson (1990, 1991) and Maugin and Epstein (1991) — see also Maugin (1993, Chap. 8).

1.2 One-dimensional motion in the Eulerian description

1.2.1 *Basic equations*

In 3D the local balances of linear momentum and energy are given by the following two equations in the absence of body forces:

$$\rho \frac{d}{dt}(\mathbf{p}/\rho) - \nabla \cdot \boldsymbol{\sigma} = \mathbf{0} \quad \text{or} \quad \frac{\partial}{\partial t}\mathbf{p} - \nabla \cdot (\boldsymbol{\sigma} - \mathbf{v} \otimes \mathbf{p}) = \mathbf{0}, \qquad (1.5)$$

and

$$\rho\frac{d}{dt}(H/\rho) - \nabla\cdot\mathbf{S} = 0 \quad \text{or} \quad \frac{\partial}{\partial t}H - \nabla\cdot(\boldsymbol{\sigma}\cdot\mathbf{v} - H\mathbf{v}) = \mathbf{0}, \qquad (1.6)$$

where

$$\mathbf{p} = \rho\mathbf{v}, \qquad H = \rho\left(\frac{1}{2}\mathbf{v}^2 + w\right), \qquad \mathbf{S} = \boldsymbol{\sigma}\cdot\mathbf{v} \qquad (1.7)$$

are the linear momentum, the total (kinetic plus potential) energy per unit actual volume, and the mechanical energy-flux (Poynting-Umov vector). Here $\boldsymbol{\sigma}$ is the (symmetric) Cauchy stress. For a volume V of regular boundary ∂V equipped with unit outward pointing normal $\mathbf{n}$, equation (1.5) yields the global balance

$$\frac{d}{dt}\mathbf{P}^t = \int_{\partial V}(\boldsymbol{\sigma} - \mathbf{p}\otimes\mathbf{v})\cdot\mathbf{n}\,ds, \qquad (1.8)$$

with

$$\mathbf{P}^t = \int_V \mathbf{p}\,dV; \qquad (1.9)$$

We consider longitudinal wave processes in an elastic thin rod seen as a 1D body. For a segment of length $L = x_2 - x_1$ of a rod, equation (1.8) yields the following time rate of change of linear momentum:

$$\frac{d}{dt}P^t = \left[\sigma - \rho v^2\right]_{x_1}^{x_2}. \qquad (1.10)$$

This is Newton's equation for the line segment with an active force

$$F = [\bar{\sigma}]_{x_1}^{x_2}, \qquad \bar{\sigma} := \sigma - \rho v^2. \qquad (1.11)$$

According to Brillouin [1925], the time average of this quantity over a time interval T defines the "force of wave pressure" or *radiative stress* by

$$\langle F\rangle := \lim_{T\to\infty}\frac{1}{T}\int_0^T F\,dt = [\langle\bar{\sigma}(x,t)\rangle]_{x_1}^{x_2}, \qquad (1.12)$$

supposing that the quantities ρ, σ and v change only under the action of a deformation wave. We see that the so defined radiative stress differs from the constant component of elastic stress $\langle\sigma\rangle$ by a value equal to *twice* the kinetic energy (this is *Leibniz's vis-viva*). It is agreed upon that the portion of the total momentum of the medium related to the deformation wave is referred to as the *wave momentum*. A more precise mathematical definition will be given in a further chapter on the basis of a variational formulation (*cf.* Ostrovsky and Potapov (1999)). What must be retained at the present stage is that the expression of the wave momentum starts with quadratic quantities in the wave amplitude. As a consequence, its time average for a travelling periodic wave does not vanish. This indicates that a longitudinal wave travelling in one direction cannot be excited in a rod if a nonzero momentum is not specified in the medium. Because of the involved nonlinearities we shall consider an approach via successive approximations.

1.2.2   Method of perturbations

For a nonlinear elastic 1D body in the *Eulerian description*, we envisage the application of a method of successive approximations in the neighbourhood of a rest unstrained (and unstressed) state by looking for various fields in the following form:

$$\rho = \rho_0 + \rho' + \rho'' + \dots, \quad u = u' + u'' + \dots, \quad \sigma = \sigma' + \sigma'' + \dots,$$
$$v = v' + v'' + \dots = u'_t + u'_x u'_t + u''_t + \dots, \tag{1.13}$$
$$e = e' + e'' = u'_x + (u'_x)^2 / 2 + u''_x + \dots$$

Although we have not introduced any ordering parameter $\varepsilon \ll 1$, it is to be understood that primed quantities are of order ε and those with a double prime are of order ε^2 with, say, $\rho'/\rho_0 \approx \varepsilon$. For the linear momentum we obtain thus

$$p = \rho\, v = \rho_0\, v' + \rho'\, v' + \rho_0\, v'' + \dots, \tag{1.14}$$

while the force of wave pressure (1.12) for a perfectly absorbing obstacle (i.e., with $\overline{\sigma}(x_2) = 0$) will yield

$$\langle F \rangle = \langle \sigma' \rangle + \langle \sigma'' \rangle - \rho_0 \langle v'^2 \rangle + \dots \tag{1.15}$$

It is assumed that the elastic constitutive equation is in the simple form (kind of limited expansion)

$$\sigma = E\,(e + \beta\, e^2), \tag{1.16}$$

where e is given by the last of (1.13), E is the linear elasticity coefficient and β stands for the nondimensional parameter of nonlinearity, respectively. For $\beta > 0$ it is said that the medium presents a *"soft"* nonlinearity, while for $\beta > 0$ the nonlinearity is said to be *"hard"*.

1.2.3   First-order approximation

This yields the wave equation

$$\frac{\partial^2 u'}{\partial t^2} - c_0^2 \frac{\partial^2 u'}{\partial x^2} = 0, \qquad c_0^2 = \frac{E}{\rho_0}, \tag{1.17}$$

while all other perturbations are deduced from u' *via* the relations

$$\rho' = -\rho_0\, u'_x, \quad v' = u'_t, \quad e = u'_x, \quad \sigma' = E\, u'_x. \tag{1.18}$$

The wave equation (1.17) has an integral of motion obtained by multiplying it by u'_x and rearranging and grouping terms, yielding the *transport of wave momentum* in the form

$$\frac{\partial}{\partial t} p^w + \frac{\partial}{\partial x} H' = 0, \tag{1.19}$$

wherein

$$p^w = -\rho_0\, u'_t\, u'_x, \qquad H' := \frac{1}{2}\left(\rho_0\, u'^2_t + E\, u'^2_x\right).\qquad(1.20)$$

Here it happens that the energy of the linear wave coincides with the density flux of wave momentum. But this is an artefact due to the 1D nature of the problem where only scalar quantities are involved. In contrast, the equation of transport of momentum (1.5) reads

$$\frac{\partial}{\partial t}\left(\rho_0\, u'_t - \rho_0\, u'_t\, u'_x\right) - \frac{\partial}{\partial x}\left(E\, u'_x - \rho_0\, u'^2_t\right) = 0.\qquad(1.21)$$

But the linear terms satisfy equation (1.17) so that the quadratic terms yield

$$\frac{\partial}{\partial t}\left(-\rho_0\, u'_t\, u'_x\right) + \frac{\partial}{\partial x}\left(\rho_0\, u'^2_t\right) = 0.\qquad(1.22)$$

This coincides with (1.19), on the average in time, because it is checked that

$$\langle \rho_0\, u'^2_t \rangle = \langle H' \rangle.\qquad(1.23)$$

For a travelling harmonic wave $u' = U\cos(k\,x - \omega\,t)$, we obtain

$$\langle p^w \rangle = \frac{1}{2}\,\rho_0\, k\,\omega\, U^2,\qquad(1.24)$$

whereas the pressure exerted on a completely absorbing obstacle is given by

$$\langle F \rangle = \frac{1}{2}\,\rho_0\,\omega^2\, U^2.\qquad(1.25)$$

On comparing (1.24) and (1.25) we obtain the remarkable result that

$$\langle F \rangle = c\,\langle p^w \rangle, \qquad c = \omega^2/k.\qquad(1.26)$$

This result relates the momentum transported by the wave and its action on a completely absorbing obstacle.

1.2.4   *Second-order approximation*

We then have to consider the following inhomogeneous equation that accounts for the nonlinear elasticity (*cf.* Andreev, 1995; Andrejew, 1940):

$$\frac{\partial^2 u''}{\partial x^2} - c_0^2\,\frac{\partial^2 u''}{\partial x^2} = \frac{\partial}{\partial x}\left[c_0^2\left(\beta - \frac{1}{2}\right)\left(\frac{\partial u'}{\partial x}\right)^2 - \left(\frac{\partial u'}{\partial t}\right)^2\right].\qquad(1.27)$$

Second-order perturbations ρ'', v'', e'' and σ'' are now expressed through the derivatives of u' and u'' by the formulas

$$\rho'' = -\rho_0\, u''_x, \qquad v'' = u''_t + u'_x\, u'_t, \tag{1.28}$$

and

$$e'' = u''_x - \frac{1}{2}\, u'^2_x, \qquad \sigma'' = E\left(u''_x + \left(\beta - \frac{1}{2}\right) u'^2_x\right). \tag{1.28_2}$$

Equation (1.27) is an equation for the linear wave $u''(x,t)$ in a medium with distributed source of external force of which the density is determined by the first-order solution $u'(x,t)$. But for formulating a closed problem in the second-order approximation, this equation must be supplemented by initial and boundary conditions. Note that the boundary conditions should be introduced already for variable values of coordinates because displacements of the boundaries also have to be of the second order.

Now, the equation of momentum transfer takes on the form

$$\frac{\partial}{\partial t}\left(\rho_0\, u'_t + \rho_0\, u''_t\right)$$
$$-\frac{\partial}{\partial x}\left[E\left(u'_x + u''_x + \left(\beta - \frac{1}{2}\right) u'^2_x\right) - \rho_0\, u'^2_t\right] = 0. \tag{1.29}$$

But accounting for equation (1.17), we deduce an inhomogeneous wave equation that completely coincides with (1.27). As to the momentum density and the wave pressure force — *cf.* (1.15) — in the second-order approximation, from (1.29) they are given by

$$p = \rho_0\, u'_t + \rho_0\, u''_t, \tag{1.30}$$

and

$$\langle F \rangle = \left[E\,\langle u'_x \rangle + E\,\langle u''_x \rangle + E\left(\beta - \frac{1}{2}\right)\langle u'^2_x \rangle - \rho_0\,\langle u'^2_t \rangle\right]_{x_1}^{x_2}. \tag{1.31}$$

This last expression exhibits a significant alteration compared to the expression of the first-order approximation. This result may be clarified by treating examples of propagation in a rod.

1.2.5 *Example of momentum and radiative stress in a thin rod*

For the sake of illustration we consider the kinematic excitation of vibrations in a rod which is rigidly fixed at $x = 0$, and with the other end with a law of motion $x(t) = l + u(t)$, with no initial perturbations. At the *first order* equation (1.17) holds true within the interval $0 \leq x \leq l$ with the conditions

$$u'|_{x=0} = 0, \qquad u'|_{x=l} = u(t), \quad t \geq 0 \tag{1.32}$$

and

$$u'|_{t=0} = u'_t|_{t=0} = 0 \tag{1.32}_2$$

For an excitation $u(t) = U_0 (1 + \cos \omega t)$ in non-resonant condition (i.e., $\omega \neq \omega_n = n \pi l c_0$) the solution to this problem may be found in books as

$$u(x,t) = U_0 \left(1 - \frac{\sin(\omega x/c)}{\sin(\omega l/c)} \cos \omega t \right)$$
$$= U_0 \left\{ 1 - \frac{1}{2 \sin(\omega l/c)} \left[\sin \omega \left(t + \frac{x}{c} \right) + \sin \omega \left(t - \frac{x}{c} \right) \right] \right\}, \tag{1.33}$$

$$u_t(x,t) = \frac{U_0 \, \omega}{\sin(\omega l/c)} \sin \frac{\omega x}{c} \sin \omega t$$
$$= \frac{U_0 \, \omega}{2 \sin(\omega l/c)} \left[\cos \omega \left(t + \frac{x}{c} \right) + \cos \omega \left(t - \frac{x}{c} \right) \right]. \tag{1.34}$$

From these one can calculate all relevant force and energy characteristic quantities of the wave process in the first approximation. In particular, the average values of wave momentum of travelling waves $u(t \pm (x/c))$ are equal in magnitude and have opposite signs:

$$\langle p_{\pm}^w \rangle = \mp \frac{1}{2c} \rho_0 \, U_0^2 \, \omega^2, \tag{1.35}$$

where the "+" sign corresponds to the wave travelling in the positive direction of the x-axis, and the "−" sign to the wave propagating in the opposite direction. Their sum corresponding to the wave momentum of the *standing wave* is equal to zero: *the standing wave does not transport momentum.* At the same time, the time average density of momentum flux

$$\langle \bar{\sigma} \rangle = -\rho_0 \langle u'^2_t \rangle + E \langle u'^2_x \rangle$$

in the standing wave is a function of the coordinate. It coincides with twice the average value $\langle K \rangle$ of the kinetic energy. Here,

$$\langle \bar{\sigma} \rangle = \langle W \rangle \sin^2 \frac{\omega x}{c} = 2 \langle K \rangle, \tag{1.36}$$

where

$$\langle W \rangle = \frac{\frac{1}{2} \rho_0 \, U_0^2 \, \omega^2}{\sin^2(\omega l/c)},$$

is the average value of the density of the total wave energy. In turn, this means that, on the average, the force of wave pressure acting on the element of the rod is given by

$$P = -\langle W \rangle \left[\sin^2 \frac{\omega\, x}{c} \right]_{x_1}^{x_2}. \qquad (1.37)$$

This expression shows that this force depends on the position relative to the nodes and crests of the standing wave.

An analysis of expression (1.37) shows that the force of pressure exerted by the wave field on the element of the medium of width $\Delta t < \lambda/4$ (where λ is the wavelength) is always directed towards decreasing momentum flux density, i.e., towards the nodes of the velocity field. This feature of wave pressure forces in liquids and gases is referred to as acoustic levitation (see, e.g., Zarembo and Krasil'nikov (1966), or Danilov and Mironov (2000)). It may be employed to "suspend" and localize particles in definite areas of the acoustic field (see Figure 1.1).

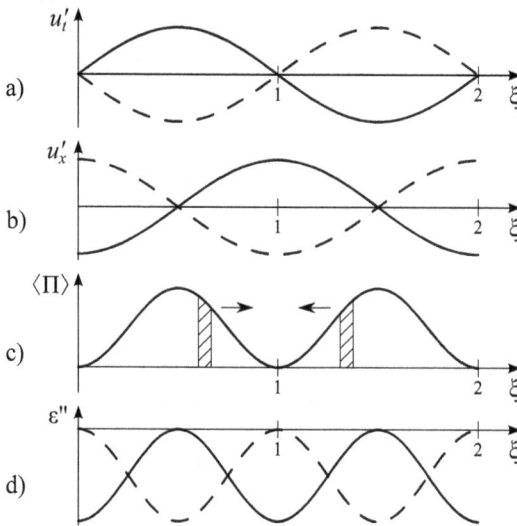

Fig. 1.1   A spatial distribution of the kinematic and power characteristics in a field of a standing wave illustrating the acoustic levitation effect: (a) speed; (b) displacement gradient; (c) averaged density of the momentum flux; (d) constant component of the strain (after Potapov *et al.*, 2005, Fig. 2).

One may inquire whether one can employ the notion of wave momentum within the framework of a linear problem neglecting terms of the same order of smallness as obtained in the second approximation of the perturbation theory? The arguments presented below show that in seeking average values of quadratic quantities it is legitimate in some cases to use the value of wave momentum calculated in the linear approximation. Account of terms in the second approximation does not contradict it. In principle the solution of the problem at the second approximation is much more complex than at the first one because in Euler variables the second boundary condition must be written at a *movable boundary* $x = l(t)$. However, some features of wave processes in the second approximation that are of interest to us can be found without solving the corresponding boundary value problem. From general considerations it follows that, in the considered situation, the time average stress and velocity of particles in the rod must be equal to zero, i.e.,

$$\langle v' \rangle = \langle v'' \rangle = 0, \qquad \langle \sigma' \rangle = \langle \sigma'' \rangle = 0. \tag{1.38}$$

On substituting from this into (1.18) and (1.28) we find that

$$\langle u'_t \rangle = 0, \qquad \langle u''_t \rangle = -\langle u'_x u'_t \rangle,$$
$$\langle u'_x \rangle = 0, \qquad \langle u''_x \rangle = -\left(\beta - \tfrac{1}{2}\right)\langle u'^2_x \rangle. \tag{1.39}$$

It follows from this that (1.30) and (1.31) yield

$$\langle p \rangle = -\rho_0 \langle u'_x u'_t \rangle = \langle p^w \rangle, \tag{1.40}$$

and

$$\langle F \rangle = -\left[\rho_0 \langle u'^2_t \rangle\right]^{x_2}_{x_1}. \tag{1.41}$$

That is, the expressions of the force of wave pressure and of the time average of the momentum density in the second approximation coincide with those in the first approximation. But allowance for the effects of the second approximation leads to the appearance of a constant component of strain (see Figure 1.1) given by

$$\langle e'' \rangle = -\beta \langle u'^2_x \rangle = -\frac{\beta \omega^2 U_0^2}{2 c_0^2} \cos^2 \frac{\omega x}{c_0}. \tag{1.42}$$

This strain is distributed in space with a period twice as small as the wavelength and its sign depends on the sign of the nonlinearity parameter. Accordingly, for a soft nonlinearity this strain is positive and the rod is elongated. For a hard nonlinearity the rod is shortened. This effect is rather similar to that of heating on a nonlinear elastic body.

1.3   One-dimensional motion in the Lagrangian description

1.3.1   *Basic equations*

Anticipating somewhat on the general presentation (*cf.* Chap. 2) we remind the reader of the equations of mass conservation and linear momentum in the Lagrangian description of space-time coordinates (X^K, t) in 3D. These equations read (compare to (1.5))

$$\rho_0 = \rho\, J_F, \qquad \frac{\partial(\rho_0\, v_i)}{\partial t} - \frac{\partial T^K_{.i}}{\partial X^K} = 0, \tag{1.43}$$

where

$$J_F = \det \mathbf{F}, \quad \mathbf{F} = \{\partial x_i / \partial X^K\}, \quad T^K_{.i} = J_F\,(\partial X^K / \partial x_j)\,\sigma_{ji}.$$

In 1D where $\mathbf{F}$ reduces to the scalar $\partial x / \partial X$ we have $J_F\, F^{-1} = 1$ so that we can rewrite the two equations in (1.43) as

$$\rho_0 = \rho\,\frac{\partial x}{\partial X}, \qquad \rho_0\,\frac{\partial^2 v}{\partial t^2} - \frac{\partial \sigma_L}{\partial X} = 0. \tag{1.44}$$

Here the Lagrangian stress scalar denoted σ_L has the same numerical value as the Eulerian stress scalar σ. The elastic constitutive equation can be proposed as

$$\sigma_L = E\left(e_L + \beta_L\, e_L^2\right), \tag{1.45}$$

where the nonlinearity parameter β_L is generally different from β and the surviving Lagrangian strain is given by the reduction

$$\mathbf{E} = \frac{1}{2}\left(\mathbf{F}^T\,\mathbf{F} - 1_R\right)$$

$$\Rightarrow e_L = \frac{1}{2}\left[\left(\frac{\partial x}{\partial X}\right)^2 - 1\right] = \frac{\partial u}{\partial X} + \frac{1}{2}\left(\frac{\partial u}{\partial X}\right)^2. \tag{1.46}$$

Because of this definition and the fact that $\sigma_L = \sigma$ a comparison between (1.45) and (1.16) allows one to relate the two parameters β and β_L.

Now we can establish the expression of the momentum and wave-pressure force for a material rod consisting of the same particles and contained between the Lagrangian coordinates X_1 and X_2. Because of the mass conservation locally written as

$$dm = \rho\, dx = \rho_0\, dX,$$

the linear momentum of the material segment can be written in the following equivalent forms:

$$P_L = \int_{X_1}^{X_2} p_L\, dX = \int_{X_1}^{X_2} \rho_0 \left(\frac{\partial u}{\partial t}\right)_{\dot{X}=\text{const}} dX = \int_{x_1(t)}^{x_2(t)} \rho\, v\, dX. \tag{1.47}$$

The distinction in definition of momenta in Eulerian and Lagrangian descriptions leads to different values of their time derivatives due to differentiation of the integrals in (1.47) with variable limits of integration, that is, as

$$\frac{dP_L}{dt} = \frac{dP}{dt} + \left[\rho\, v^2\right]_{x_1(t)}^{x_2(t)}. \tag{1.48}$$

From this it follows that the forces acting on the rod segment in both descriptions are related by

$$F_L = F + \left[\rho\, v^2\right]_{x_1(t)}^{x_2(t)}, \tag{1.49}$$

and from this, on account of (1.11),

$$\left[\sigma_L\right]_{X_1}^{X_2} = \left[\sigma\right]_{x_1(t)}^{x_2(t)}, \tag{1.50}$$

so that we can say that the elastic stresses σ_L and σ acting in the same segment of the medium are equal. For the wave-pressure force, from (1.49) we directly obtain

$$\langle F_L \rangle = \langle F \rangle + \left[\langle \rho\, v^2 \rangle\right]_{x_1(t)}^{x_2(t)}. \tag{1.51}$$

The difference appears only when quantities of the second order of smallness are taken into account. In conclusion of this point we note that the time variation of the momentum p_L is definitively related to mass transport, while the force F_L is determined by the difference in elastic Lagrangian stresses at the boundary of the stationary element $\Delta X = X_2 - X_1$.

1.3.2   *Perturbation analysis at the first-order of approximation*

At the first order the displacement $u'(X,t)$ is governed by the linear wave equation

$$\frac{\partial^2 u'}{\partial t^2} - c_0^2\, \frac{\partial^2 u'}{\partial X^2} = 0. \tag{1.52}$$

Perturbations in the other variables are then given by

$$\rho' = -\rho_0\, u'_X, \qquad e'_L = u'_X, \qquad \sigma'_L = E\, u'_X.$$

It follows that

$$p'_L = \rho_0\, u'_X, \qquad P'_L = [E\, u'_X]_{X_1}^{X_2}. \tag{1.53}$$

Unlike the analogous quantities in the Eulerian description they are linear in the wave amplitude, and their time average *vanishes* for periodic processes.

1.3.3 Perturbation analysis at the second order of approximation

The relevant wave equation is obtained as

$$\frac{\partial^2 u''}{\partial t^2} - c_0^2 \frac{\partial^2 u''}{\partial X^2} = \left(\beta_L + \frac{1}{2}\right) \frac{\partial}{\partial X}(u_X'^2), \tag{1.54}$$

and the other variables are deduced from the relations

$$\rho'' = -\rho_0\, u_X'' + \rho_0\, u_X'^2, \qquad e_L'' = u_L'' + \frac{1}{2}\, u_X'^2, \tag{1.55}$$

and

$$\sigma_L'' = E\left(u_X'' + \left(\beta_L + \frac{1}{2}\right) u_X'^2\right). \tag{1.56}$$

Note that the first of (1.55) already differs from the corresponding continuity equation in Eulerian coordinates. As to the momentum density and the wave-pressure force in the second approximation, they are given by the following expressions:

$$p_L = \rho_0\, u_t' + \rho_0\, u_t'', \tag{1.57}$$

and

$$\langle F_L \rangle = E\left[\langle u_X'' \rangle + \left(\beta_L + \frac{1}{2}\right) \langle u_X'^2 \rangle\right]_{X_1}^{X_2}. \tag{1.58}$$

Clearly, the Lagrangian momentum is given by the same expression as in the Eulerian description, while the wave-pressure force is equal to the difference of Lagrangian stresses at the boundary of the segment ΔX. In order to comprehend the similarities and differences between the two descriptions it is necessary to compare results for the *same* wave problem, namely, the kinematic excitation of vibrations already examined in Section 1.2. With one end fixed at $X = 0$ and the excitation at the other end $X = l$, at the first order of approximation one has to solve a problem with field equation (1.52) in the interval $0 \le X \le l$ and with boundary conditions and initial conditions

$$u'|_{X=0} = 0, \qquad u'|_{X=l} = U(t), \quad t \ge 0, \tag{1.59}$$

and

$$u'|_{t=0} = u_t'|_{t=0} = 0. \tag{1.60}$$

This is equivalent to the problem already stated in the Eulerian description. The solution is given by equations (1.32) provided we replace x by X. But we must still note that, as distinct from the Eulerian description, in the Lagrangian formalism there is no definition of wave momentum analogous to the first of (1.20), and the average value of the momentum flux vanishes both for travelling and standing waves. Of course, considering a "definition" of wave momentum in the form of $(1.20)_1$ (i.e. $p_L^w := -\rho_0 \, u_X' \, u_t'$) and of a corresponding equation for its transport would be quite reasonable from a strict mathematical viewpoint, but the physical meaning would be doubtful. Interesting results will ensue only when quadratic quantities are taken into account, hence going to the second-order approximation. We may question whether second-order quantities will lead to informative results. Again, from general considerations there must be no nonzero time average values of both particle velocity and stresses. That is,

$$\langle u_t' \rangle = \langle u_t'' \rangle = 0, \qquad \langle \sigma_L' \rangle = \langle \sigma_L'' \rangle = 0. \tag{1.61}$$

This provides

$$\langle u_X' \rangle = 0, \qquad \langle u_X'' \rangle = -E \left(\beta_L + \frac{1}{2} \right) \langle u_X'^2 \rangle. \tag{1.62}$$

Here, whereas the linear quantities coincide with those of the the Eulerian approach, those of in the second order of approximation differ from those of the Eulerian description. But it also follows that $\langle p_L \rangle = 0$ since no transport of mass occurs. But F_L is also zero. This not so much surprising as in the Eulerian description the corresponding quantities are nonzero because of the presence of convective terms that are naturally absent in the Lagrangian description. However, a nonzero constant strain appears in the rod in the latter description. It is given by

$$\langle e_L'' \rangle = -\beta_L \langle u_X'^2 \rangle. \tag{1.63}$$

This is formally the same as the expression given in equation (1.42).

Another problem example for a rod with rigidly fastened ends is reported in Potapov *et al.* (2005). A more rigorous analysis of nonlinear longitudinal waves in a rod of finite length was also previously given by Miloserdova and Potapov (1983).

1.4 Summary and concluding remarks

The following general traits can be gathered from the foregoing brief developments.

First we must emphasize the important differences between Eulerian and Lagrangian descriptions. In defining the physical meaning of one or another theoretical result in continuum mechanics one should proceed from the fact that an observer (experimentalist) is in a *laboratory frame* of reference that consists of a time (Newtonian) and spatial (Eulerian) coordinate system. Therefore, the quantities calculated in Eulerian variables are regarded to be observable (i.e., true) physical quantities. Thus, the wave pressure force and the wave momentum defined in Eulerian coordinates should be considered as physically meaningful variables. If one consistently keeps to these definitions of radiative stress and wave transported momentum, then the same physical results will be obtained, independently of the method of description of the wave field. In Eulerian variables the wave pressure force is defined as a time average of the medium momentum flux across the boundaries of the volume fixed in space. It differs from the wave pressure force (1.51) in Lagrangian variables by a quadratic value of the convective component of momentum flux across the boundary of the volume.

The Lagrangian coordinate system is "frozen-in" the medium and evolves during the body motion. Physical quantities defined in these variables are conventional to a certain extent because the values of parameters of the state of the medium at a current instant of time are compared with the values at the initial instant, when the body was undeformed (natural state), rather than with the values at the preceding time instant. Furthermore, the physical interpretation of theoretical calculations depends on specific formulations of the problem. The differences between calculations according to Euler and Lagrange are quantities of the second order of smallness in the wave amplitude and arise because the first ones are calculated for a material volume fixed in space, and the others for an individual mass volume the value and spatial shape of which are constantly changing in time. As a consequence, a sufficiently rigorous mathematical definition of the notion of wave momentum is possible only at the first approximation in Eulerian variables. It is a quadratic component of the body momentum related to variations of the medium density in the wave field $(1.20)_1$, whereas no notion of wave momentum exists in Lagrangian variables. Wave momentum no longer has a rigorous mathematical definition in the second approximation. But its notion can be successfully used in a number of problems that require calculation of averaged values, for instance, when there is no time averaged body motion.

Finally, allowance for *anharmonicity* leads to the appearance in an elastic medium of nonzero, time averaged deformations and stresses generated by the wave field, that are related to the physical nonlinearity of the material only. In the second approximation they are described by similar expressions both in Eulerian and Lagrangian variables.

The above conclusions pertain to the case of acoustics in elastic solids. But the notion of wave momentum has been the object of warm discussions and vivid debates mostly in the fields of fluid mechanics, optics and electromagnetism. The most thorough analysis of this notion was probably given in a paper by McIntyre (1981), a specialist of atmospheric flows. His standpoint is that the notion of "wave momentum" is a *myth* essentially because wave momentum is not related to a transport of mass. But what really matters is the *flux of momentum* because in Brillouin's own words (*cf.* Brillouin (1925), "the latter may very well differ from zero even when the density of momentum is zero" (translation by McIntyre from the French original). McIntyre relates the original "mistake" of Rayleigh in 1905 in words telling that "if the reflexion of a train of waves exercises a pressure upon a reflector, it can only be because the train of waves itself involves momentum. Of course, in standard continuum mechanics, the *flux of momentum* — none other than the (Cauchy) stress — can perfectly well exist without there being any momentum present. However, from the point of view of *field theory*, if such a force as the radiative pressure on an obstacle exists, it is legitimate to ask from what "conservation law" this force follows and what is, therefore, the corresponding quantity to be conserved or not conserved from a dynamical view point?

The wave momentum seems to be the proper candidate to play this role. If it is not so much physical, then it deserves its name of pseudo-momentum. This qualification was used by physicists, in particular Peierls (1976, 1979, 1985, 1991) in so far as the propagation of photons — the grains of light — is concerned. It is the thought experiment envisaged by Peierls in his examination of the recall of a prism (the Jones-Richards experiment) by light that kindled the interest of one of the authors (G.A.M), because this thought experiment appeared just like a copy of a reasoning introduced some thirty years before by J. D. Eshelby in solid mechanics — and reproduced in Maugin (1993, pp. 24–27 and the footnote in p. 209), — to evaluate the "force" acting on a material inhomogeneity — i.e., the force dual, from an energetic viewpoint, of the observed displacement of the inhomogenity (a region of foreign material) on the material manifold upon mechanical loading of the material specimen. This force is not of Newtonian character; it is of thermodynamic nature. Going then to dynamics, it was found that it was indeed a *pseudo*-momentum that was the associated dynamic quantity (*cf.* Maugin and Trimarco (1992)) in a field-theoretic vision in which Noether's fundamental theorem about conservation laws was central. This "pseudo-momentum" could be defined by transport back (so-called "pull-back") to the material manifold of the traditional physical linear momentum. The non-Newtonian nature of the inhomogeneity force and the associated momentum were considered by these authors (G.A.M and Trimarco) from a very pragmatic point of view: they can be exploited in a post-processing strategy to build criteria of progress in the theory of structural defects. We are far from Brillouin's original thinking, although one easily realizes that the presence of an obstacle or the border of a domain manifests a material inhomogeneity in the overall considered problem.

The relevant discussion in problems of wave propagation in fluid dynamics is thoroughly reported in McIntyre (1981) with plenty of spot on references — see also Andrew and McIntyre (1978) and Stone (2000). In the case of electromagnetism in matter the peak of discussions was attained in the period 1970–1990 with enlightening contributions by many authors among whom Blount (1971) and Nelson (1979) must be underlined. We agree with the wise viewpoint reached by Nelson (1990, 1991) that physical momentum, pseudo-momentum and field (wave) momentum (the latter according to Brenig (1955)) are all different. We refer the reader to the detailed discussion reported in Maugin (1993, Chaps. 8 and 9). In all we gather that there exist circumstances where pseudo-momentum is conserved while physical momentum is not. This is the case for light particles referred to as photons. For instance (*cf.* Peierls (1985)), pseudo-momentum of photons is conserved in refraction or reflection of light while physical momentum is not conserved for photons for an object completely immersed in a refractive medium as is the case in the crucial experiment of Jones and Richards (1954) and Jones and Leslie (1978). The analysis, albeit brief, given in the present chapter fully reveals the importance of distinguishing between Eulerian and Lagrangian descriptions. In what follows we shall heavily rely on the so-called material framework of deformable solids, and the general scheme of invariance and related conservation laws provided by field-theoretical arguments will be privileged with a pragmatic vision in so far as applications of the notion of wave momentum is concerned.

Remark 1.2. Herein above the average used is one over time and may more carefully be noted $\langle .. \rangle_T$: thus (1.24) can be rewritten as

$$\langle p^w \rangle_T = \frac{1}{2} \rho_0 \, k \, \omega \, U^2 = \frac{1}{\lambda} \rho_0 \, \pi \, \omega \, U^2. \qquad (a)$$

In certain works and further chapters, in order to associate quasi-particle motion with the wave momentum, we are led to introduce the sum over one wavelength λ of the wave momentum. Dividing the result by λ yields a space average that should be noted $\langle .. \rangle_S$. It is checked that this agrees with (a), i.e.

$$\langle p^w \rangle_T = \langle p^w \rangle_S. \tag{b}$$

Chapter 2

Elements of continuum thermomechanics

Object of the chapter

This chapter introduces all elements of the nonlinear thermo-mechanics of continua needed in the rest of this book. The presentation is somewhat traditional, avoiding unnecessary abstraction and geometric formalism so as to be accessible to mechanicians as well as physicists, acousticians, and applied mathematicians. General balance laws and theorems of thermodynamics are paid special attention with both Eulerian (spatial) and material descriptions in so-called Euler–Cauchy and Piola–Kirchhoff formats. Elasticity is introduced in finite- and small- strain pictures for the benefit of nonlinear and linear physical acoustics.

Note. This is not a course on the general thermomechanics of continua. Such more comprehensive presentations can be found in now classical books such as Truesdell and Toupin (1960), and Eringen (1967) — also Maugin (1988, 2011b). We avoid any advanced element of differential geometry. Only material needed for a good understanding of the sequel is introduced. However, general lines for the formulation of thermodynamically admissible constitutive equations are given for further use.

2.1 Material body

The **motion** (or deformation mapping) of the material body B of $M = M^3$ — the set of material points X — is the *time ordered sequence* $C(X)$ of the positions, sometimes called *placements*, occupied by the point X in Euclidean *physical space* E^3, the arena of classical phenomenological physics. This is expressed by the sufficiently (as needed) regular space-

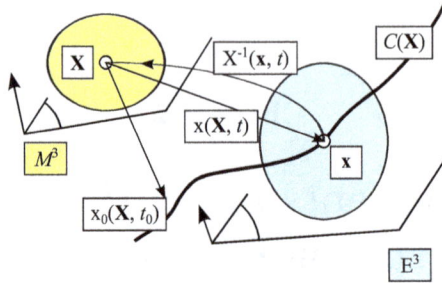

Fig. 2.1   The classical view of finite transformations (direct and inverse motions).

time parametrized mapping (*cf.* Figure 2.1)

$$\mathbf{x} = \overline{\mathbf{x}}(\mathbf{X}, t). \tag{2.1}$$

This is often (but not necessarily) reported to a Cartesian system of coordinates x^i, $i = 1, 2, 3$. The set of geometrical points $\mathbf{x}(B, t$ fixed) constitutes the *actual* or *current configuration* K_t of the body at time t. Usually, an origin of time, say t_0, is chosen such that $t_0 < t$, and (2.1) then reads

$$\mathbf{x}_0 = \overline{\mathbf{x}}(\mathbf{X}, t_0). \tag{2.2}$$

When both mappings (2.1) and (2.2) are sufficiently smooth, and in particular, invertible, we can rewrite (2.1) as

$$\mathbf{x} = \overline{\mathbf{x}}\left(\overline{\mathbf{x}}^{-1}(\mathbf{x}_0, t_0), t\right) = \overline{\mathbf{x}}(\mathbf{x}_0, t; t_0) = \widetilde{\mathbf{x}}(\mathbf{x}_0, t). \tag{2.3}$$

This representation of the direct motion is called Lagrangian, the $\mathbf{x}_0$ being Lagrangian coordinates. The configuration $K_0 = K_t(t = t_0)$ of the body, the *initial configuration* at $t = t_0$, belongs to the sequence of "actual" configurations. This is the motion description preferred in fluid mechanics. Many authors identify the two representations (2.1) and (2.3) by identifying $\mathbf{X}$ and $\mathbf{x}_0$. But the motion representation (2.1) is somewhat more abstract and is essentially due to Gabrio Piola (1848) in a paper of far reaching insight. While (2.1) is called *the direct motion mapping* $K_R \to K_t$ *at* t, in the same smoothness conditions as above, *the inverse* motion is given by

$$\mathbf{X} = \overline{\mathbf{x}}^{-1}(\mathbf{x}, t). \tag{2.4}$$

The direct $\mathbf{F}$ and inverse $\mathbf{F}^{-1}$ motion **gradients** are defined thus

$$\mathbf{F} := \nabla_R \overline{\mathbf{x}} \equiv \frac{\partial \overline{\mathbf{x}}}{\partial \mathbf{X}}, \qquad \mathbf{F}^{-1} := \nabla \overline{\mathbf{x}}^{-1} \equiv \frac{\partial \overline{\mathbf{x}}^{-1}}{\partial \mathbf{x}}. \tag{2.5}$$

It is immediately checked that

$$\mathbf{F} \cdot \mathbf{F}^{-1} = \mathbf{1}, \qquad \mathbf{F}^{-1} \cdot \mathbf{F} = \mathbf{1}_R, \qquad (2.6)$$

where the symbols $\mathbf{1}$ and $\mathbf{1}_R$ represent the unit dyadics in E^3 and on M, respectively. It must be emphasized that both $\mathbf{F}$ and $\mathbf{F}^{-1}$ are *not* tensors in the traditional sense because they are geometric objects defined on two different manifolds simultaneously. In a picturesque language, we can say that they have one foot in K_t and another in K_R. Such objects are so-called *two-point tensor fields*. They have components

$$\mathbf{F} = \left\{ F_K^i \equiv F_{iK} \right\}, \qquad \mathbf{F}^{-1} = \left\{ \left(\mathbf{F}^{-1}\right)_i^K = \left(\mathbf{F}^{-1}\right)^{Ki} \right\}, \qquad (2.7)$$

where the upward or downward position of lower Latin indices is irrelevant by virtue of the Cartesian representation chosen in K_t. Speaking of an *a priori* symmetry of $\mathbf{F}$ and $\mathbf{F}^{-1}$ is a mathematical nonsense since one must specify with respect to what metric is tensorial symmetry defined. The Jacobian determinant of $\mathbf{F}$ is noted

$$J_F = \det \mathbf{F}. \qquad (2.8)$$

Of course,

$$J_{F^{-1}} = \det \mathbf{F}^{-1} = \left(J_F\right)^{-1}. \qquad (2.9)$$

Matter density

If ρ_0, the matter density at $\mathbf{X}$, does not depend on time, the actual mass density ρ is related to ρ_0 by the change of volume between configurations, i.e.,

$$\rho(\mathbf{x}, t) = \rho_0(\mathbf{X}) \, J_F^{-1}. \qquad (2.10)$$

Since densities are always positive, only deformation mappings such that J_F *is positive and never vanishes*, are considered. In physical terms this signifies the *impenetrability* of matter.

Displacement field

This is the field $\mathbf{u}(\mathbf{X}, t)$ or $\overline{\mathbf{u}}(\mathbf{x}, t)$ defined by

$$\mathbf{x} = \mathbf{X} + \mathbf{u}(\mathbf{X}, t) \qquad \text{or} \qquad \mathbf{x} = \mathbf{X} + \overline{\mathbf{u}}(\mathbf{x}, t). \qquad (2.11)$$

On taking the material gradient ∇_R of the first of these and the spatial gradient ∇ of the second we obtain with (2.5),

$$\mathbf{F} = \mathbf{1} + \mathbf{H}, \quad \mathbf{H} \equiv \nabla_R \mathbf{u}; \qquad \mathbf{F}^{-1} = \mathbf{1} - \mathbf{h}, \quad \mathbf{h} \equiv \nabla \overline{\mathbf{u}}. \qquad (2.12)$$

Pull-back and push-forward operations

They are tensorial transformations effected with the help of the motion mapping itself since these operations are conducted between two different manifolds. Historically first, but also endowed with a definite relevance in continuum mechanics, is the convection operation introduced by G. Piola, the *Piola transformation*. Let $\mathbf{A}$ be a vector field in the actual configuration. Then the material contravector defined by

$$\overline{\mathbf{A}} = J_F \, \mathbf{F}^{-1} \, \mathbf{A} = \left\{ \overline{A}^K = J_F \left(\mathbf{F}^{-1} \right)^{Ki} A_i \right\} \tag{2.13}$$

is the *Piola transform* of $\mathbf{A}$. Conversely,

$$\mathbf{A} = J_F^{-1} \, \mathbf{F} \, \overline{\mathbf{A}} = \left\{ A^i = J_F^{-1} \, F_K^i \, \overline{A}^K \right\} \tag{2.14}$$

In this connection the reader will note the following two demonstrable identities:

$$\nabla_R \left(J_F \, \mathbf{F}^{-1} \right) = \mathbf{0}, \qquad \nabla \left(J_F^{-1} \, \mathbf{F} \right) = \mathbf{0}, \tag{2.15}$$

from which there follows that

$$\nabla_R \cdot \overline{\mathbf{A}} = J_F \, \mathbf{F} \cdot \left(\nabla_R \mathbf{A} \right) = J_F \, \nabla \cdot \mathbf{A}. \tag{2.16}$$

Of course, this reminds us of the formula for the change of elementary volume, dv and dV, between the actual and reference configurations:

$$dv = J_F \, dV, \tag{2.17}$$

so that

$$\left(\nabla_R \cdot \overline{\mathbf{A}} \right) dV = \left(\nabla \cdot \mathbf{A} \right) dv. \tag{2.18}$$

By the same token it is salient to remind the reader of the so-called Nanson's formula for the change between oriented surface elements $\mathbf{n} \, da$ and $\mathbf{N} \, dA$ of the same surface with respective unit normals $\mathbf{n}$ and $\mathbf{N}$ in the actual and reference configurations:

$$\mathbf{n} \, da = J_F \, \mathbf{N} \cdot \mathbf{F}^{-1} \, dA, \qquad \mathbf{N} \, dA = J_F^{-1} \, \mathbf{n} \cdot \mathbf{F} \, da. \tag{2.19}$$

Matter velocity

Motion, as compared to statics and equilibrium, has to do with the *time* evolution of fields, whereas most of the operations mentioned before had to do with the spatial variation of fields. In this regard the basic notion in continuum mechanics is that of velocity field. From the two descriptions

(2.1) and (2.4) of the motion, we can define the *physical* velocity field $\mathbf{v}$ in K_t and the *material* velocity field $\mathbf{V}$ in K_R by

$$\mathbf{v}(\mathbf{X}, t) := \left.\frac{\partial \overline{\mathbf{x}}}{\partial t}\right|_X, \qquad \mathbf{V}(\mathbf{x}, t) := \left.\frac{\partial \overline{\mathbf{x}}^{-1}}{\partial t}\right|_x \tag{2.20}$$

Please note the functional dependence indicated in equations (2.20). We let the reader check by way of exercise the following two important relations between these two velocities:

$$\mathbf{v} + \mathbf{F}\,\mathbf{V} = \mathbf{0}, \qquad \mathbf{V} + \mathbf{F}^{-1}\,\mathbf{v} = \mathbf{0}. \tag{2.21}$$

The derivatives used in (2.20) are called the material (or Lagrangian) time derivative, and the spatial (Eulerian) time derivative since, depending on the case, they are taken at fixed material coordinates (particle) or fixed spatial coordinates. We clearly have the following operational definitions (prove these by way of exercise)

$$\begin{aligned}
\left.\frac{\partial}{\partial t}\right|_{X \text{ fixed}} &= \left.\frac{\partial}{\partial t}\right|_{x \text{ fixed}} + \mathbf{v} \cdot \nabla, \\
\left.\frac{\partial}{\partial t}\right|_{x \text{ fixed}} &= \left.\frac{\partial}{\partial t}\right|_{X \text{ fixed}} + \mathbf{V} \cdot \nabla_R,
\end{aligned} \tag{2.22}$$

since we verify with (2.21) that

$$\mathbf{v} \cdot \nabla = -\mathbf{V} \cdot \nabla_R. \tag{2.23}$$

2.2 Balance laws of the thermomechanics of continua

2.2.1 *Global balance laws in the Euler–Cauchy format*

It has become an established tradition in modern continuum physics to introduce the basic balance laws of continuum mechanics and thermodynamics in *global* form, i.e., for a *whole* deformable body B. This can be achieved in two different formalisms — which are not fully independent of one another —, whether the postulate of these balance laws is made in the actual configuration K_t or on the material manifold (configuration K_R). To that purpose we note B_R and ∂B_R the open simply connected region occupied by the body in K_R and its regular boundary (in principle without edges and apices) with uniquely defined unit outward pointing normal $\mathbf{N}$, and B_t and ∂B_t, the image of these in the actual configuration, with uniquely defined outward pointing normal $\mathbf{n}$, it being understood that the body still is simply connected in this configuration, that is, no holes have formed during the deformation mapping. The material bodies considered in this section are classical continua exhibiting no visible evolving microstructure, and they are acted upon at most by body forces and surface forces. From the thermodynamic viewpoint, they may conduct heat and dissipate energy. They may also present body sources of energy and admit a flux of energy at their boundaries. They are submitted to the first and second laws of thermodynamics. We are thus led to introducing the following quantities:

- the mass density $\rho(\mathbf{x}, t)$ in K_t and $\rho_0(\mathbf{X}, t)$ in K_R,
- the body force $\mathbf{f}(\mathbf{x}, t)$ per unit mass in K_t,
- an applied traction $\mathbf{T}^d(\mathbf{x}, t; \mathbf{n})$ at ∂B_t,
- an internal energy e per unit mass in K_t,
- a body source of energy $h(\mathbf{x}, t)$ per unit mass in K_t,
- an energy influx $q(\mathbf{x}, t; \mathbf{n})$ at ∂B_t,
- an entropy η per unit mass in K_t,
- a body source of entropy $\hat{\eta}$ per unit mass in K_t,
- an entropy influx $s(\mathbf{x}, t; \mathbf{n})$ at K_t,
- an internal source of entropy $\tilde{\eta}$ per unit mass in B_t.

Exchanging no mass with its exterior and no mass being created internally in this modelling, the material body is endowed with an invariant mass measure dm, such that

$$dm = \rho \, dv = \rho_0 \, dV = \text{const.} \tag{2.24}$$

Then the following *global balance laws* hold true in the actual configuration (so-called *Eulerian* representation and *Cauchy* format) of these laws:

- *Balance of mass*:

$$\frac{d}{dt} \int_{B_t} dm = 0, \tag{2.25}$$

- *Balance of linear (physical) momentum*:

$$\frac{d}{dt} \int_{B_t} \mathbf{v} \, dm = \int_{B_t} \mathbf{f} \, dm + \int_{\partial B_t} \mathbf{T}^d \, da; \tag{2.26}$$

- *Balance of angular (physical) momentum*:

$$\frac{d}{dt} \int_{B_t} \mathbf{x} \times \mathbf{v} \, dm = \int_{B_t} \mathbf{x} \times \mathbf{f} \, dm + \int_{\partial B_t} \mathbf{x} \times \mathbf{T}^d \, da; \tag{2.27}$$

- *First law of thermodynamics*:

$$\frac{d}{dt} \int_{B_t} (H^t/\rho) \, dm = P_{\text{ext}}(B_t) + \int_{B_t} h \, dm - \int_{\partial B_t} q \, da; \tag{2.28}$$

- *Balance of entropy*:

$$\frac{d}{dt} \int_{B_t} \eta \, dm - \int_{B_t} \hat{\eta} \, dm + \int_{\partial B_t} s \, da = \int_{B_t} \tilde{\eta} \, dm; \qquad (2.29)$$

- *Second law of thermodynamics*:

$$\int_{B_t} \tilde{\eta} \, dm \geq 0. \qquad (2.30)$$

Here we have defined the energy per unit volume or Hamiltonian density, the kinetic energy E^t, and the power expanded by external source P_{ext} by

$$H^t = K^t + E^t, \qquad K^t = \frac{1}{2} \rho(\mathbf{x}, t) \, \mathbf{v}^2, \qquad E^t = \rho \, e, \qquad (2.31)$$

and

$$P_{\text{ext}}(B_t) = \int_{B_t} \mathbf{f} \cdot \mathbf{v} \, dm + \int_{\partial B_t} \mathbf{T}^d \cdot \mathbf{v} \, da. \qquad (2.32)$$

Note that the balance law (2.27) contains no new notion, e.g., there is no applied couple per unit volume in this modelling.

Following a celebrated lemma due to Cauchy (*cf.* Truesdell and Toupin (1960)), the surface data $(\mathbf{T}^d, q, s)$ are linear affine in the unit normal $\mathbf{n}$. Accordingly, we can write

$$\mathbf{T}^d(\mathbf{x}, t; \mathbf{n}) = \mathbf{n} \cdot \boldsymbol{\sigma}, \qquad q(\mathbf{x}, t; \mathbf{n}) = \mathbf{q} \cdot \mathbf{n}, \qquad s(\mathbf{x}, t; \mathbf{n}) = \mathbf{n} \cdot \mathbf{s}, \qquad (2.33)$$

where the second-order (spatial) tensor $\boldsymbol{\sigma}$ is called the *Cauchy stress*, $\mathbf{q}$ is the (spatial) *heat* (in) *flux vector*, and $\mathbf{s}$ is the (spatial) *entropy* (in) *flux vector*. Equations (2.33) mean that the "internal forces" introduced and their thermal analogues depend at most on the first-order description of the geometry of the limiting surface. This applies to any facet cut in the body. Accordingly, we are not considering here any generalized continuum mechanics (which call for the notions of hyperstresses, couple stresses, *etc.*).

2.2.2 Euler–Cauchy format of the local balance laws of thermomechanics

Because of the invariance of dm and the fact that dV and B_R are material, it is immediately checked that we have the following obvious results for any P:

$$\frac{d}{dt} \int_{B_t} P \, dm = \int_{B_t} \frac{dP}{dt} \, dm, \qquad \frac{d}{dt} \int_{B_R} P \, dV = \int_{B_R} \left. \frac{\partial P}{\partial t} \right|_X dV. \qquad (2.34)$$

On using the first of these, the divergence theorem where it applies, and enforcing the basic working continuity hypothesis of continuum physics according to which the final integrands must vanish for any spatial volume element and surface element, we obtain the following local balance laws from equations (2.25) through (2.30):

- *Balance of mass, also called the continuity equation:*

$$\dot{\rho} + \rho(\nabla \cdot \mathbf{v}) = \left. \frac{\partial \rho}{\partial t} \right|_x + \nabla \cdot (\mathbf{p}^t) = 0, \qquad \mathbf{p}^t := \rho \mathbf{v}; \qquad (2.35)$$

- *Balance of linear (physical) momentum:*

$$\rho \dot{\mathbf{v}} - div\, \boldsymbol{\sigma} = \rho \mathbf{f}, \qquad (2.36a)$$

  or

$$\left. \frac{\partial}{\partial t} \mathbf{p}^t \right|_x + \nabla \cdot (\mathbf{p}^t \otimes \mathbf{v} - \boldsymbol{\sigma}) = \rho \mathbf{f}; \qquad (2.36b)$$

- *Balance of angular (physical) momentum:*

$$\boldsymbol{\sigma} = \boldsymbol{\sigma}^T \text{ i.e., } \sigma_{ij} = \sigma_{ji} \text{ or } \sigma_{[ij]} = 0; \qquad (2.37)$$

- *First law of thermodynamics:*

$$\rho \frac{d}{dt}(H^t/\rho) - \nabla \cdot (\boldsymbol{\sigma} \cdot \mathbf{v} - \mathbf{q}) = \rho\,(h + \mathbf{f} \cdot \mathbf{v}), \qquad (2.38a)$$

  or

$$\left. \frac{\partial}{\partial t} H^t \right|_x + \nabla \cdot (\mathbf{p}^t\,(H^t/\rho) - \boldsymbol{\sigma} \cdot \mathbf{v} + \mathbf{q}) = \rho\,(h + \mathbf{f} \cdot \mathbf{v}); \qquad (2.38b)$$

- *Balance of entropy:*

$$\rho \frac{d}{dt}(S^t/\rho) + \nabla \cdot \mathbf{s} - \rho\,\hat{\eta} = \rho\,\tilde{\eta}, \qquad (2.39a)$$

  or

$$\left. \frac{\partial}{\partial t} S^t \right|_x + \nabla \cdot (\mathbf{p}^t\,\eta + \mathbf{s}) - \rho\,\hat{\eta} = \rho\,\tilde{\eta}; \qquad (2.39b)$$

- *Second law of thermodynamics:*

$$\rho\,\tilde{\eta} \geq 0. \qquad (2.40)$$

Equations (2.37) mean that the Cauchy stress is symmetric in the absence of microstructure and body couple. Equations (2.35), (2.36b), (2.38b) and (2.39b) take the form of strict conservation laws in the spatial framework in the absence of any source.

2.2.3 Global balance laws in the Piola–Kirchhoff format

Global balance laws can also be postulated for material domains over M. To that purpose, we introduce the Piola transformation (*cf.* equation (2.13)) of several thermomechanical fields (stresses, heat flux, and entropy flux):

$$\mathbf{T} = J_F \, \mathbf{F}^{-1} \boldsymbol{\sigma}, \quad \mathbf{Q} = J_F \, \mathbf{F}^{-1} \mathbf{q}, \quad \mathbf{S} = J_F \, \mathbf{F}^{-1} \mathbf{s}. \tag{2.41}$$

The first of these defines the *first Piola–Kirchhoff stress* (*not* a usual second-order tensor because the transformation is performed only on the first index, so that the transformation is only partial, yielding a two-point tensor field). The second and third of (2.41) define the *material* heat and entropy fluxes. Then the postulate of the basic balance laws of thermomechanics in the Piola–Kirchhoff form is given by the following set of equations:

$$\frac{d}{dt} \int_{B_R} dm = 0, \tag{2.42}$$

$$\frac{d}{dt} \int_{B_R} \mathbf{p}_R \, dV = \int_{B_R} \rho_0 \, \mathbf{f} \, dV + \int_{\partial B_R} \mathbf{N} \cdot \mathbf{T} \, dA, \tag{2.43}$$

$$\frac{d}{dt} \int_{B_R} H_R \, dV = \int_{B_R} \rho_0 \, (h + \mathbf{f} \cdot \mathbf{v}) \, dV - \int_{\partial B_R} \mathbf{N} \cdot (\mathbf{Q} - \mathbf{T} \cdot \mathbf{v}) \, dA, \tag{2.44}$$

$$\frac{d}{dt} \int_{B_R} S_R \, dV - \int_{B_R} \rho_0 \, \tilde{\eta} \, dV + \int_{\partial B_R} \mathbf{N} \cdot \mathbf{S} \, dA = \int_{B_R} \rho_0 \, \tilde{\eta} \, dV \geq 0. \tag{2.45}$$

Here we have not recalled the balance of angular momentum which is a secondary notion, and we have defined a few new quantities by

$$\mathbf{p}_R = \rho_0 \, \mathbf{v}, \qquad \mathbf{p}^t = \rho \, \mathbf{v} = J_F^{-1} \, \mathbf{p}_R,$$

$$H_R = K^R + E^R, \quad K^R = \frac{1}{2} \, \rho_0 \, \mathbf{v}^2, \tag{2.46}$$

$$E^R = \rho_0 \, e, \qquad S_R = \rho_0 \, \eta.$$

2.2.4 Piola–Kirchhoff format of the local balance laws of thermomechanics

In full parallelism with what was done for the Euler–Cauchy format, on using the second of (2.34), the material divergence theorem where it applies, and enforcing the basic working hypothesis of continuum physics according to which the final integrands must vanish for any material volume element and surface element, we obtain the following local balance laws from equations (2.42) through (2.45):

- *Balance of mass, also called the continuity equation:*

$$\frac{\partial}{\partial t}\rho_0 \bigg|_X = 0; \tag{2.47}$$

- *Balance of linear (physical) momentum:*

$$\frac{\partial}{\partial t}\mathbf{p}_R \bigg|_X - div_R \, \mathbf{T} = \rho_0 \, \mathbf{f}; \tag{2.48}$$

- *Balance of angular (physical) momentum:*

$$\mathbf{F}\,\mathbf{T} = \mathbf{T}^T \, \mathbf{F}^T; \tag{2.49}$$

- *First law of thermodynamics:*

$$\frac{\partial}{\partial t}H_R \bigg|_X - \nabla_R \cdot (\mathbf{T} \cdot \mathbf{v} - \mathbf{Q}) = \rho_0 \, \mathbf{f} \cdot \mathbf{v}; \tag{2.50}$$

- *Balance of entropy:*

$$\frac{\partial}{\partial t}S_R \bigg|_X + \nabla_R \cdot \mathbf{S} - \rho_0 \, \hat{\eta} = \Sigma_R := \rho_0 \, \tilde{\eta}; \tag{2.51}$$

- *Second law of thermodynamics:*

$$\Sigma_R \geq 0. \tag{2.52}$$

These equations are not independent of those deduced previously in the spatial frame. In particular, (2.49) is just a rewriting of (2.37) in the material framework. As to (2.48) — *which still has components in physical space* — it is readily shown to follow from (2.36) by multiplying the latter by J_F, and exploiting (2.9) and (2.31). Again, equations (2.48), (2.49), (2.50) and (2.51) take the form of strict conservation laws in the material formalism $(\mathbf{X}, t)$ in the absence of source terms. However, sometimes, these equations are referred to as "material" equations, as compared to the spatial equations deduced in the Cauchy format. This is a misnomer because only the space-time parametrization and partial derivatives here refer to this framework while both equations (2.48) and (2.49) still have components in the physical framework (actual configuration K_t). We shall see in the next chapter how one constructs equations that are completely in the material framework, both in terms of tensor objects, and space-time parametrization.

2.3   General theorems of thermodynamics

2.3.1   *Thermodynamic hypotheses*

We denote by $\theta > 0$, $\inf \theta = 0$, the thermodynamic temperature. Such a notion as well as that of entropy are well defined only in thermostatics (see any book on thermodynamics, e.g., Maugin (1999a)) where θ is given by

$$\theta = \frac{\partial \bar{e}(., \eta)}{\partial \eta} > 0, \qquad e = \bar{e}(., \eta), \qquad (2.53)$$

where the missing argument may be any other variable of state such as strain in thermoelasticity. It is traditional to introduce another thermodynamic function, the *Helmholtz free energy* ψ, by the following Legendre transformation

$$e + (-\psi) = \eta\,\theta, \qquad \eta = \frac{\partial(-\psi)}{\partial \theta} = -\frac{\partial \psi}{\partial \theta}. \qquad (2.54)$$

The first of these equations places e and ψ in duality in the sense of convex analysis. Since the Legendre transformation conserves convexity, if, as it should, the internal energy e is *convex* in η, then the free energy ψ is *concave* in θ, so that the specific heat C_θ is always positive:

$$C_\theta = -\theta \frac{\partial^2 \psi}{\partial \theta^2} > 0. \qquad (2.55)$$

Modern thermodynamics, in the manner of Coleman and Noll (*cf.* Truesdell and Noll (1965)), assumes that the notions of temperature and entropy are still defined in true thermo-*dynamics*, that is, outside thermodynamic equilibrium. We shall assume the same as also the fact that heat body source and heat flux and entropy body source and entropy flux are related by

$$\hat{\eta} = h/\theta, \qquad \mathbf{s} = \mathbf{q}/\theta, \qquad \mathbf{S} = \mathbf{Q}/\theta. \qquad (2.56)$$

2.3.2   *Local expression of the general theorems of thermomechanics*

We apply the above-enunciated working hypotheses. Then we can establish the following theorems (here tr = trace).

- *Kinetic energy theorem*:
  On taking the inner product of (2.36a) with $\mathbf{v}$, we obtain that

$$\rho\,\frac{d(K^t/\rho)}{dt} - \nabla \cdot (\boldsymbol{\sigma} \cdot \mathbf{v}) + tr\left(\boldsymbol{\sigma} \cdot (\nabla \mathbf{v})^T\right) = \rho\,\mathbf{f} \cdot \mathbf{v}; \qquad (2.57)$$

- *Internal energy theorem*:

  On expanding (2.38a) and combining with (2.57) we obtain that

$$\rho \frac{d(E^t/\rho)}{dt} - tr\left(\boldsymbol{\sigma} \cdot (\nabla \mathbf{v})^T\right) + \nabla \cdot \mathbf{q} = \rho\, h; \qquad (2.58)$$

- *Clausius–Duhem inequality*:

  Combining now (2.58) and (2.38a) and introducing the free energy density $\psi = e - \eta\,\theta$, we obtain the following inequality:

$$-\rho\,(\dot{\psi} + \eta\,\dot{\theta}) + tr\left(\boldsymbol{\sigma} \cdot (\nabla \mathbf{v})^T\right) - (\mathbf{q}/\theta) \cdot \nabla\theta = \rho\,\theta\,\tilde{\eta} \geq 0, \quad (2.59)$$

while (2.39a) takes also the alternate form

$$\rho\,\theta\,\dot{\eta} + \nabla \cdot \mathbf{s} = \rho\,h + \rho\,\theta\,\tilde{\eta}. \qquad (2.60)$$

Ultimately, equation (2.60) is the equation that will govern the temperature field.

In direct parallelism with these spatial equations, it is easy to establish the following equations in the *Piola–Kirchhoff* form:

- *Kinetic energy theorem*:

$$\left.\frac{\partial K_R}{\partial t}\right|_X - \nabla_R \cdot (\mathbf{T} \cdot \mathbf{v}) + tr\left(\mathbf{T} \cdot (\nabla_R \mathbf{v})^T\right) = \rho_0\, \mathbf{f} \cdot \mathbf{v}; \qquad (2.61)$$

- *Internal energy theorem*:

$$\left.\frac{\partial E_R}{\partial t}\right|_X - tr\left(\mathbf{T} \cdot (\nabla_R \mathbf{v})^T\right) + \nabla_R \cdot \mathbf{Q} = \rho_0\, h; \qquad (2.62)$$

- *Clausius–Duhem inequality*:

$$\theta\,\Sigma_R = -\left(\dot{W} + S\,\dot{\theta}\right) + tr\left(\mathbf{T} \cdot (\nabla_R \mathbf{v})^T\right) - (\mathbf{Q}/\theta)\,\nabla_R\theta \geq 0, \quad (2.63)$$

where we introduced the free energy $W = E^t - S\,\theta$, per unit volume in the reference configuration.

- *Heat propagation equation*:

$$\theta\,\dot{S} + \nabla_R \cdot \mathbf{S} = \rho_0\, h + \theta\,\Sigma_R. \qquad (2.64)$$

2.4   Finite-strain elasticity

Elasticity is the prime focus of this book. In anthropomorphic words, it can be said that elasticity reflects a material behavior of the nondissipative type which is summarized as: the material remembers only one state, the initial unloaded state. Following Green, the corresponding constitutive equation can be derived from a potential energy. The main ingredient in the formulation is that of *strain*.

2.4.1 *Measures of finite strains*

Deformation measures are typical "metrics" (truly symmetric tensors). Some of them can be defined thus (here the superscript T denotes the operation of *transposition*, δ's are Kronecker symbols):

$$\mathbf{C}(\mathbf{X},t) := \mathbf{F}^T\,\mathbf{F} = \left\{ C_{KL} = F_K^i\,\delta_{ij}\,F_L^j \right\}, \tag{2.65}$$

$$\mathbf{C}^{-1} := (\mathbf{F}^{-1})\,(\mathbf{F}^{-1})^T = \left\{ (\mathbf{C}^{-1})^{KL} = (\mathbf{F}^{-1})_i^K\,\delta^{ij}\,(\mathbf{F}^{-1})_j^L \right\}. \tag{2.66}$$

These two measures, defined over M, are called the *Cauchy–Green* finite (material) strain tensor and the *Piola* finite (material) strain tensor, respectively. They are the inverse to one another. However, these measures are *absolute* ones. They are not compared to an undeformed metric. A natural *relative* strain measure called the Lagrangian strain is given by

$$\mathbf{E} := \frac{1}{2}\,(\mathbf{C} - \mathbf{1}_R). \tag{2.67}$$

2.4.2 *Time rates of finite strains*

We easily compute

$$\left.\frac{\partial \mathbf{F}}{\partial t}\right|_X = \mathbf{L}\,\mathbf{F}, \tag{2.68}$$

where the spatial rate $\mathbf{L}$ — velocity-gradient tensor — is given by

$$\mathbf{L} = \dot{\mathbf{F}}\,\mathbf{F}^{-1} = (\nabla \mathbf{v})^T = \{\nu_{i,j}\}. \tag{2.69}$$

Here a superimposed dot is used as an alternate notation for the material time derivative.

The symmetric part of $\mathbf{L}$ is the *strain-rate* (spatial) tensor given by

$$\mathbf{D} = \mathbf{L}_S = \frac{1}{2}\left((\nabla \mathbf{v})^T + \nabla \mathbf{v} \right), \tag{2.70}$$

where the subscript S denotes the operations of symmetrization. We could as well define strain rates in material space, e.g. $\dot{\mathbf{E}}$. Indeed, we easily check that

$$\mathbf{D} = \mathbf{F}^{-T}\,\dot{\mathbf{E}}\,\mathbf{F}^{-1}, \qquad \dot{\mathbf{E}} = \mathbf{F}^T\,\mathbf{D}\,\mathbf{F}. \tag{2.71}$$

Here the symbolism "$-T$" denotes the transposed of the inverse. The following formulas are easily established:

$$\left.\frac{\partial J_F}{\partial t}\right|_X = J_F\,tr\,\mathbf{L} = J_F\,(\nabla \cdot \mathbf{v}), \qquad \left.\frac{\partial (dv)}{\partial t}\right|_X = (\nabla \cdot \mathbf{v})\,dv, \tag{2.72}$$

since $tr\,\mathbf{L} = tr\,\mathbf{D} = \nabla \cdot \mathbf{v}$ and $\partial(dV)/\partial t|_X = 0$.

2.4.3   *Rigid-body motions*

This special class of motions is defined in geometry by Killing's equations
for isometries (conservation of a metric in time), here, for instance,

$$\left.\frac{\partial \mathbf{E}(\mathbf{X}, t)}{\partial t}\right|_X = \mathbf{0}. \tag{2.73}$$

For non-zero $\mathbf{F}$, this is equivalent to the spatial expression

$$\mathbf{D}(\mathbf{x}, t) = \mathbf{0}, \quad \text{or } \nu_{i,j} + \nu_{j,i} = 0. \tag{2.74}$$

2.5   Strains in small-strain elasticity

In general we have the following *exact* formulas in finite-strain elasticity.

$$\begin{aligned}
\mathbf{E} &= \frac{1}{2}\left(\mathbf{H} + \mathbf{H}^T + \mathbf{H}^T \mathbf{H}\right), \\
\mathbf{e} &:= \mathbf{F}^{-T}\mathbf{E}\,\mathbf{F}^{-1} = \frac{1}{2}\left(\mathbf{h} + \mathbf{h}^T - \mathbf{h}^T \mathbf{h}\right).
\end{aligned} \tag{2.75}$$

In *small-strain theory* for which $\mathbf{H}$ and $\mathbf{h}$ are small in the sense that $|\mathbf{H}| \equiv$ (trace $\mathbf{H}^T \mathbf{H})^{1/2}$ or $|\mathbf{h}| \equiv$ (trace $\mathbf{h}^T \mathbf{h})^{1/2}$ is considered as an infinitesimal quantity of the first order, then neglecting terms of second order in the "small" displacement gradients, we obtain the following approximation

$$\mathbf{E} = \mathbf{e} = (\nabla \mathbf{u})_S \equiv \frac{1}{2}\left(\nabla \mathbf{u} + (\nabla \mathbf{u})^T\right). \tag{2.76}$$

The true tensor $\mathbf{e}$ — sometimes noted $\boldsymbol{\varepsilon}$ — here is called the *infinitesimal strain*. In Cartesian components in the actual configuration it has components

$$e_{ij} = u_{(i,j)} \equiv \frac{1}{2}\left(u_{i,j} + u_{j,i}\right). \tag{2.77}$$

The reader will have noticed that in this small-strain approximation we no longer distinguish between material and spatial representations. To the same degree of approximation,

$$J_F \cong 1 + tr\,\mathbf{e}, \qquad J_F^{-1} \cong 1 - tr\,\mathbf{e}. \tag{2.78}$$

Remark 2.1. In the small-strain theory where the superimposed dot may be used for any of the material and spatial time derivatives, we have the following reductions:

$$\mathbf{L} = \dot{\mathbf{F}} = \dot{\mathbf{H}} \qquad \text{or } \nu_{i,j} = \dot{u}_{i,j}, \tag{2.79}$$

$$\mathbf{D} = \dot{\mathbf{e}} \qquad \text{or } D_{ij} = \dot{e}_{ij} = \dot{u}_{(i,j)}, \tag{2.80}$$

Remark 2.2. The above given introduction to the motion and kinematics of continua is necessary and sufficient for the purpose of this book. Of course, there exist more rigorous geometric approaches involving abstract manifolds, tangent spaces, fiber-bundles, *etc.* In this line we recommend the now classic book by Marsden and Hughes (1975), as also the book of Ciarlet (1988). Along the present line we recommend the books of Eringen (1980) and Ogden (1984) with which we fully agree, and, of course, the classical treatises of Truesdell and Toupin (1960) and Truesdell and Noll (1965), the textbooks of Spencer (1976) and Chadwick (1976), the series of books on Continuum Physics edited by Eringen (1971–1976), and the books of Maugin (1988, 1993, 2011b), and Eringen and Maugin (1990a).

2.6   Constitutive equations for finite-strain elasticity

In agreement with the above-given definition of elasticity, we discard all dissipative processes and temperature effects. We need no longer distinguish between internal and free energy, and simply call $\psi = e$ and $E = W$ the elastic potential energy per unit actual mass and unit reference volume, respectively. Equations (2.58) and (2.63) then reduce to the following equations:

$$\rho\dot{\psi} = tr\left(\boldsymbol{\sigma} \cdot (\nabla\mathbf{v})^T\right) = tr(\boldsymbol{\sigma} : \mathbf{D}) = \boldsymbol{\sigma} : \mathbf{D}, \tag{2.81}$$

and

$$\left.\frac{\partial W}{\partial t}\right|_X = tr\left(\mathbf{T} \cdot (\nabla_R\mathbf{v})^T\right). \tag{2.82}$$

For a general, possibly *anisotropic* and *inhomogeneous*, elastic body the energy density can be considered in the following forms

$$W = \overline{W}(\mathbf{F}; \mathbf{X}) = \widehat{W}(\mathbf{E}; \mathbf{X}), \tag{2.83}$$

where the second expression is invariant with respect to time-dependent rotations of the actual frame, i.e., is *objective*. More specific dependences of W for any crystallographic groups can be found in Eringen and Maugin (1990a, Vol. I, Appendix). $\widehat{W}$ is supposed to be sufficiently smooth, and convex in its argument $\mathbf{E}$. On computing the material time derivative of W, we have

$$\dot{W} = tr\left(\frac{\partial\overline{W}}{\partial\mathbf{F}} \cdot (\nabla_R\mathbf{v})^T\right) = tr\left(\frac{\partial\widehat{W}}{\partial\mathbf{E}} \cdot \dot{\mathbf{E}}\right). \tag{2.84}$$

Accordingly, for all motions that are not those of a rigid body (*cf.* equation (2.50)), equation (2.82) yields the elastic constitutive equation

$$\mathbf{T} = \frac{\partial \overline{W}}{\partial \mathbf{F}}; \qquad \mathbf{S} = \frac{\partial \widehat{W}}{\partial \mathbf{E}} = \mathbf{T} \cdot \mathbf{F}^{-T}. \qquad (2.85)$$

The second of these is called the *second Piola–Kirchhoff stress* or *energy stress*, this one a symmetric true material (contravariant) tensor. We immediately have a symmetric Cauchy stress given by

$$\boldsymbol{\sigma} = J_F^{-1} \, \mathbf{F} \cdot \mathbf{T} = J_F^{-1} \, \mathbf{F} \cdot \mathbf{S} \cdot \mathbf{F}^T. \qquad (2.86)$$

Another mixed material stress of possible interest is the so-called *Mandel stress* defined by

$$\mathbf{M} := \mathbf{T} \cdot \mathbf{F} = \mathbf{S} \cdot \mathbf{C}, \qquad (2.87)$$

which may be said to be symmetric with respect to $\mathbf{C}$ — considered as a deformable material metric — since it is checked that

$$\mathbf{C} \cdot \mathbf{M} = \mathbf{C} \cdot \mathbf{S} \cdot \mathbf{C} = (\mathbf{C} \cdot \mathbf{M})^T, \qquad (2.88)$$

by virtue of the symmetry of $\mathbf{S}$.

In problems of *nonlinear crystal acoustics*, the linear momentum equation (2.48) is therefore written in components as

$$\frac{\partial}{\partial X^K} \left[S^{KL} \left(\delta_{iL} + u_{i,L} \right) \right] + \rho_0 \, f_i = \rho_0 \, \frac{\partial^2 u_i}{\partial t^2}, \qquad (2.89)$$

with $\mathbf{S}$ given by the second of (2.85).

In the *isotropic* case, according to a well known representation theorem of isotropic scalar-valued functions of a symmetric tensor due to Cauchy, the functional dependence of $\widehat{W}$ on tensor $\mathbf{E}$ reduces to a dependence on the three basic scalar invariants of this tensor, i.e.,

$$W = \widehat{W}(E_\alpha, \ \alpha = 1, 2, 3; \mathbf{X}) \qquad (2.90)$$

with

$$E_1 = tr\,\mathbf{E}, \quad E_2 = tr\,\mathbf{E}^2, \quad E_3 = tr\,\mathbf{E}^3. \qquad (2.91)$$

Of course, the equations for finite strains are useful in problems involving directly finite strains such as in rubber-like and polymeric materials as also biological materials such as soft tissues where the two conditions of isotropy and incompressibility are more than often enforced. But they are also a necessary point of departure in the study of small motions superimposed on a finite deformation.

2.7 Constitutive equations for small-strain elasticity

Here we consider small strains in the absence of an initially finitely deformed or stressed state. This is the material behaviour most often considered by mechanical engineers (in the isotropic and homogeneous case) and also in linear crystal acoustics (anisotropic case). In the latter case the Cauchy stress reduces to the linear representation as

$$\sigma_{ij} = E_{ijkl}\, e_{kl}, \tag{2.92}$$

with a tensor of elasticity coefficients E_{ijkl} given by

$$E_{ijkl} = E_{klij} = E_{(ij)(kl)} := \frac{\partial^2 \widehat{W}}{\partial E_{MN}\, \partial E_{PQ}}\, \delta_{iM}\, \delta_{jN}\, \delta_{Pk}\, \delta_{Ql}. \tag{2.93}$$

In the case of isotropic linear elasticity, only quadratic terms at most are kept in the representation of the potential energy, which therefore reads

$$W = \frac{1}{2}\,\lambda\,(tr\,\mathbf{e})^2 + \mu\,tr(\mathbf{e}^2), \tag{2.94}$$

where λ and μ are the celebrated Lamé coefficients of elasticity. This corresponds to

$$E_{ijkl} = \lambda\,\delta_{ij}\,\delta_{kl} + \mu\,(\delta_{ik}\,\delta_{jl} + \delta_{il}\,\delta_{ik}), \tag{2.95}$$

and

$$\sigma_{ij} = \lambda\,e_{kk}\,\delta_{ij} + 2\,\mu\,e_{ij}. \tag{2.96}$$

Positive definiteness of the energy requires that the Lamé coefficients satisfy the following inequalities:

$$3\,\lambda + 2\,\mu \geq 0, \qquad \mu \geq 0. \tag{2.97}$$

The Young modulus E, the Poisson ratio ν, and the bulk modulus B are then defined by

$$E = \frac{\mu\,(3\,\lambda + 2\,\mu)}{\lambda + \mu}, \qquad \nu = \frac{\lambda}{2\,(\lambda + \mu)}, \qquad B = \frac{3\,\lambda + 2\,\mu}{3}. \tag{2.98}$$

Coefficient μ is also called the *shear* modulus.

Chapter 3

Pseudomomentum and Eshelby stress

Object of the chapter

Here, the notions of pseudomomentum and Eshelby stress are introduced on the basis of the fundamental balance equations formulated in Chapter 2 for finite- and small-strains. These quantities are related to the notion of conservation laws that consider the whole physical system and not degrees of freedom separately. This can be obtained by direct manipulations effected on the balance laws or by application of the celebrated Noether's theorem of field theory in the presence of a variational formulation. The notion of wave momentum emerges from such general considerations. Special features of the one-dimensional case are emphasized, while the dissipative case is shown to be handled by mimicking Noether's identity. Noether's theorem is recalled for all practical purposes.

3.1 Introduction

It was shown in Chapter 2 that for a standard continuum we have the following basic local equations of continuity, linear (physical) momentum, moment of momentum and energy in the Eulerian formulation (in the absence of heat processes and body couple):

$$\dot{\rho} + \rho\left(\nabla \cdot \mathbf{v}\right) = \left.\frac{\partial \rho}{\partial t}\right|_x + \nabla \cdot (\mathbf{p}^t) = 0, \qquad \mathbf{p}^t := \rho\mathbf{v}, \qquad (3.1)$$

$$\left.\frac{\partial}{\partial t}\mathbf{p}^t\right|_x + \nabla \cdot (\mathbf{p}^t \otimes \mathbf{v} - \boldsymbol{\sigma}) = \rho\mathbf{f}, \qquad (3.2)$$

$$\boldsymbol{\sigma} = \boldsymbol{\sigma}^T \quad \text{i.e.,} \quad \sigma_{ij} = \sigma_{ji} \quad \text{or} \quad \sigma_{[ij]} = 0, \qquad (3.3)$$

$$\rho\frac{d}{dt}(H^t/\rho) - \nabla \cdot (\boldsymbol{\sigma} \cdot \mathbf{v}) = \rho\left(h + \mathbf{f} \cdot \mathbf{v}\right), \qquad (3.4)$$

to which must be added boundary conditions at the actual boundary in physical space where data can be prescribed at all times, and initial-value data at all points for a time-evolution problem.

We note that Equations (3.2) and (3.3) have components in a local actual frame (in the language of physicists, laboratory frame) in physical space, while all derivatives involved are with respect to the space-time coordinates (x^i, t). For further progress the following question can be raised: *Is it possible to express the above system entirely on the material manifold with all partial derivatives with respect to the space-time parametrization* (X^K, t)? In particular, is there some way to replace (3.2) by an equation of conservation of the form

$$\left. \frac{\partial \mathbf{P}}{\partial t} \right|_X - div_R \, \mathbf{b} = \mathbf{f}_R, \qquad (3.5)$$

where momentum $\mathbf{P}$ (a material co-vector), (a fully material) stress tensor $\mathbf{b}$ (of yet *unknown* symmetry), and a (material) force $\mathbf{f}_R$ per unit reference volume remain to be defined? We shall see in Section 3.2 that this definition in terms of the quantities present in Equations (3.1) through (3.4) is possible in particular in the case of elasticity — even though in finite deformations.

Remark 3.1. From the view point of analytical mechanics (e.g. in a formulation of the type of the principle of virtual power), a linear momentum equation such as (3.2) is generated by an infinitesimal dis-*placement* $\delta \mathbf{x}$ of components δx^i because the relevant field is the placement itself. In general it is a totally different matter with an equation such as (3.5) which would clearly be generated by an infinitesimal change of "particle" $\delta \mathbf{X}$ of components δX^K, because all fields present in this new formulation, whether of mechanical origin (e.g., placement) or else (e.g., electromagnetic fields, evolving microstructure), are parametrized by the couple (X^K, t). This search for a "material" conservation law of momentum must include all fields present in the theory and not only the translational displacement of standard continuum mechanics, because, like the energy equation, this equation concerns the *whole physical* body system under consideration. Thus, to the possible surprise of many readers, in a complex medium exhibiting an internal degree of freedom of rotation (e.g., a Cosserat deformable solid) — which usually makes that the Cauchy stress is no longer symmetric — this additional rotational effect must also contribute to the definition of both $\mathbf{P}$ and $\mathbf{b}$. This is richly documented in a recent book on so-called configurational forces (*cf.* Maugin (2011b)). For instance, in an electromagnetic body for which the body forces per unit actual volume can be shown to be expressible as (*cf.* Eringen and Maugin (1990a))

$$\mathbf{f}^{em} = div\, \mathbf{t}^{em} - \left.\frac{\partial \mathbf{p}^{em}}{\partial t}\right|_x, \tag{3.6}$$

where the expression of both $\mathbf{t}^{em}$ and $\mathbf{p}^{em}$ was formally evaluated in terms of the electromagnetic fields, it can be shown that (3.5) will be replaced by

$$\left.\frac{\partial \mathbf{P}^{tot}}{\partial t}\right|_X - div_R\, \mathbf{b}^{tot} = \mathbf{f}_R, \tag{3.7}$$

where both $\mathbf{P}^{tot}$ and $\mathbf{b}^{tot}$ have contributions from both mechanical (deformation) fields *and* electromagnetic fields. Such examples will appear in further chapters of this book when dealing with waves in particular cases of crystals.

Remark 3.2. There will remain to obtain a result concerning the possible symmetry or asymmetry of $\mathbf{b}$ or $\mathbf{b}^{tot}$, as also to answer the question whether sensible, exploitable, boundary conditions apply to these material tensors.

3.2 Pseudomomentum in hyperelastic materials

We consider here the pure mechanical case of elasticity in finite strains with constitutive equations given by (2.85)–(2.86) and envisage the construction of Equation (3.5). A partial step towards the answer to the question asked in Section 3.1 was already taken in Paragraph 2.2.4 *via* the so-called Piola–Kirchhoff format of the field equations (2.47)–(2.50), where a partial pull-back to the reference configuration was implemented. It suffices now to complete this pull-back by applying the deformation gradient $\mathbf{F}$ to the right of all terms in Equation (2.48) while taking account of the general expression of the energy density for a possibly inhomogeneous elastic material, e.g., $W = \overline{W}(\mathbf{F}; \mathbf{X})$ with $\rho_0 = \bar{\rho}_0(\mathbf{X})$. We note the following useful intermediate results:

$$(div_R\, \mathbf{T}) \cdot \mathbf{F} = div_R(\mathbf{T} \cdot \mathbf{F}) - \mathbf{T} \cdot (\nabla_R \mathbf{F})^T$$

$$= div_R(\mathbf{T} \cdot \mathbf{F}) - \nabla_R W + \left.\frac{\partial \overline{W}}{\partial \mathbf{X}}\right|_{\mathbf{F}\ \text{fixed}},$$

and

$$\left.\frac{\partial(\rho_0\, \mathbf{v})}{\partial t}\right|_X \cdot \mathbf{F} = -\left.\frac{\partial \mathbf{P}}{\partial t}\right|_X - \nabla_R K^R + \left(\frac{K^R}{\rho_0}\right)\nabla_R \rho_0,$$

since

$$\nabla_R W = \frac{\partial \overline{W}}{\partial \mathbf{F}} \cdot (\nabla_R \mathbf{F})^T + \frac{\partial \overline{W}}{\partial \mathbf{X}}\bigg|_{\mathbf{F}\text{ fixed}}.$$

We let the reader check that the result of the whole manipulation is the following equation of (material) momentum:

$$\frac{\partial \mathbf{P}}{\partial t}\bigg|_X - div_R \, \mathbf{b} = \mathbf{f}_R, \tag{3.8}$$

in which the following fully material quantities have been defined:

$$\mathbf{P} := -\mathbf{p}^t \cdot \mathbf{F} = \rho_0 \, \mathbf{C} \cdot \mathbf{V}, \tag{3.9}$$

$$\mathbf{b} := -(L\,\mathbf{1}_R + \mathbf{T} \cdot \mathbf{F}), \quad L := K^R - W, \tag{3.10}$$

$$\mathbf{f}_R = -\rho_0 \, \mathbf{f} \cdot \mathbf{F} + \mathbf{f}^{\text{inh}}, \tag{3.11}$$

with

$$\mathbf{f}^{\text{inh}} := \frac{\partial L}{\partial \mathbf{X}}\bigg|_{\text{expl}} \equiv \frac{\partial L}{\partial \mathbf{X}}\bigg|_{\mathbf{F},\mathbf{v}\text{ fixed}} = \left(\frac{K^R}{\rho_0}\right)\nabla_R \bar{\rho}_0 - \frac{\partial \overline{W}}{\partial \mathbf{X}}\bigg|_{\text{expl}}. \tag{3.12}$$

The fields $\mathbf{P}$, $\mathbf{b}$ and $\mathbf{f}^{\text{inh}}$ are referred to as the material momentum or *pseudomomentum*, the *Eshelby material stress*, and the material *force of inhomogeneity*, respectively. The latter vanishes identically in a materially homogeneous elastic body. In this condition and *in the absence of body force* $\mathbf{f}$, equation (3.8) reduces to the strict conservation law

$$\frac{\partial \mathbf{P}}{\partial t}\bigg|_X - div_R \, \mathbf{b} = \mathbf{0}. \tag{3.13}$$

Tensor $\mathbf{b}$ is a mixed covariant-contravariant second-order tensor. We let the reader check that with the symmetry condition (2.37) or (2.49), this $\mathbf{b}$ satisfies a condition of symmetry with respect to the deformed metric $\mathbf{C}$ in the form

$$\mathbf{C} \cdot \mathbf{b} = (\mathbf{C} \cdot \mathbf{b})^T. \tag{3.14}$$

This, of course, is the same symmetry as that of the Mandel stress — *cf.* Equation (2.88) — since $\mathbf{b}$ and $\mathbf{M}$ differ only by an isotropic contribution.

This fully material format is completed by noting that the continuity equation still reads as (2.47), i.e.,

$$\frac{\partial}{\partial t}\rho_0\bigg|_X = 0, \tag{3.15}$$

while the local energy equation (in the absence of heat processes and body forces) can be rewritten as

$$\frac{\partial H}{\partial t}\bigg|_X - \nabla_R \cdot ((\mathbf{b} + L\,\mathbf{1}_R) \cdot \mathbf{V}) = 0, \tag{3.16}$$

where $\mathbf{V}$ is the material velocity defined in (2.20) or (2.21).

As to the boundary condition to be verified by $\mathbf{b}$ it is of no interest in direct problem solving as it cannot be expressed only in terms of data. We have therefore answered the query proposed in Section 3.1.

Remark 3.3. The obtained full expression of $\mathbf{P}$ here is expanded as

$$\mathbf{P} = -\rho_0 \, \mathbf{v} \cdot \mathbf{F} = -\rho_0 \, \mathbf{v} \cdot \left(\mathbf{1} + (\nabla \mathbf{u})^T\right) = -\mathbf{p}^t \cdot \boldsymbol{\delta}_R - \rho_0 \, \mathbf{u} \cdot (\nabla_R \mathbf{u})^T, \quad (3.17)$$

where $\boldsymbol{\delta}_R$ is a "shifter" (with components $\delta_K^i = \delta_{iK}$) from current to material configuration, and the second contribution looks like a three-dimensional version of the *wave momentum* discussed in Chapter 1.

Remark 3.4. The scalar L defined in the second of (3.10) has the form of a *Lagrangian density* in analytical continuum mechanics (kinetic energy minus potential energy). This hints at a variational Hamilton–Lagrange formulation of the present theory as in fact evidenced in the next section.

3.3   Field-theoretical formulation in the case of elasticity

For the sake of illustration we consider a direct-motion formulation of the Hamiltonian–Lagrangian type for classical elasticity. In a rather general manner we could consider the following Lagrangian density per unit volume of K_R

$$L = \overline{L}(\mathbf{v}, \mathbf{F}; \mathbf{X}, \mathbf{x}, t) = \overline{K} - \overline{W} - \Phi_R, \quad (3.18)$$

where

$$\overline{K}(\mathbf{v}; \mathbf{X}, t) = \frac{1}{2} \rho_0(\mathbf{X}, t) \, \mathbf{v}^2,$$
$$W = \overline{W}(\mathbf{F}; \mathbf{X}, t), \qquad \Phi_R = \rho_0(\mathbf{X}, t) \, \varphi(\mathbf{x}, t). \quad (3.19)$$

This is indeed quite general because it includes a potential for *physical* body forces, *via* Φ_R, where the potential φ per unit mass is necessarily dependent on the current placement (e.g., gravity), while both the reference matter density and the elasticity potential remain dependent on both the material point $\mathbf{X}$ and Newtonian time t. Dependence on $\mathbf{X}$ means material inhomogeneity, i.e., whenever

$$\left.\frac{\partial \rho_0}{\partial \mathbf{X}}\right|_t \neq \mathbf{0}, \qquad \left.\frac{\partial \overline{W}}{\partial \mathbf{X}}\right|_{\text{expl}} \neq \mathbf{0}. \quad (3.20)$$

Explicit dependence on time t is a much rarer occurrence in books on continuum mechanics. Indeed the condition (*cf.* Epstein and Maugin (2000))

$$\left.\frac{\partial \rho_0}{\partial t}\right|_X \neq 0, \tag{3.21}$$

would mean that the reference density may evolve in time by addition or subtraction of matter by some means. This would account for the phenomenon of *growth* or *resorption* of matter, a case not considered in this book but of actuality in the biomechanics of soft tissues. As to the condition (*cf.* Maugin (2009))

$$\left.\frac{\partial \overline{W}}{\partial t}\right|_{\text{expl}} \neq 0, \tag{3.22}$$

it would mean a possible evolution in time of elastic properties, hence the phenomenon of *ageing*. This is seldom considered in view of the large time scales involved in the process for normal physical conditions for inert matter. From the point of view of analytical mechanics, systems in which the Lagrangian density depends explicitly on time are said to be *rheonomic*, according to a classification originally due to Boltzmann (*cf.* Lanczos (1962)). In contrast, systems in which L does not depend explicitly on time are called *scleronomic* systems. Most systems considered in this book are scleronomic.

A last remark concerning the Lagrangian density for the direct-motion description is that neither the kinetic energy nor the elasticity potential energy involve the *field* itself $\mathbf{x}$ — in fact its three components —, but only its space and time derivative in the parametrization $(\mathbf{X}, t)$. The reason is that elasticity is a *gauge theory* as *Galilean invariance* (translation in physical space) eliminates this possible dependence on $\mathbf{x}$ (or displacement $\mathbf{u}$ in small strains).

In order to account for both natural boundary conditions at the boundary ∂B and initial conditions (this specifies the physical momentum, position being $\mathbf{X}$ at $t = 0$), we envisage a Hamiltonian–Lagrangian variational principle in the following form:

$$\delta_X \int_0^t\!\!\int_{B_R} L \, dV \, dt + \int_0^t\!\!\int_{\partial B} \mathbf{T}^d \cdot \delta_X \, \mathbf{x} \, da \, dt$$
$$- \int_{B_R} \big((\mathbf{p}_R(t) - \mathbf{p}_R(0)) \cdot \delta_X \, \mathbf{x} \, dV = 0, \tag{3.23}$$

where the surface term corresponding to the datum of a physical traction is expressed at the deformed (actual) boundary. The variation effected in (3.23) to the price of simple calculations immediately yields the *field equation* — in fact three scalar equations — which is none other than the Piola–Kirchhoff form of the local balance of linear (physical) momentum (*cf.* (2.48)):

$$\frac{\partial \mathbf{p}_R}{\partial t}\bigg|_X - div_R \,\mathbf{T} = \rho_0 \,\mathbf{f} \quad \text{in } B_R,\tag{3.24}$$

$$\mathbf{N} \cdot \mathbf{T} = \left(\frac{da}{dA}\right) \mathbf{T}^d = \overline{\mathbf{T}}^d \quad \text{at } \partial B_R,\tag{3.25}$$

$$\mathbf{p}_R(t=0) = \mathbf{p}_R(0), \qquad \mathbf{p}_R(t=t) = \mathbf{p}_R(t),\tag{3.26}$$

of which the last one is a mere identity. Here

$$\mathbf{p}_R = \rho_0 \,\mathbf{v}, \qquad \mathbf{T} = \frac{\partial \overline{W}}{\partial \mathbf{F}}, \qquad \mathbf{f} = -\nabla \varphi,\tag{3.27}$$

and da/dA is the ratio of elementary surface elements in actual and reference configuration.

Now in the spirit of field theory we contemplate only the first contribution in (3.23) and apply to it the celebrated Noether's theorem of field theory. This theorem, proved by the German mathematician Emmy Noether (1918), probably is the most powerful theorem of mathematical physics in the twentieth century. Its quintessential content is that it associates with the *field equations* (obtained as the Euler–Lagrange equations for each independent fields — in the present case the equations of motion (3.24)) a set of *conservation laws*, each of which itself representative of the invariance (or lack of invariance) of the considered physical system under a group of transformations. In the present case of classical mechanics, we consider only invariance (or lack of invariance) under translations of the time and space parametrization $(t, X^K; K = 1, 2, 3)$. Noether's theorem and Noether's identity are recalled in the Appendix A. On applying Noether's identity to the Lagrangian of (3.23), we can write down the relevant accompanying conservation laws:

- Conservation of energy (for time translations):

$$\frac{\partial H}{\partial t}\bigg|_X - \nabla_R \cdot \mathbf{Q} = h,\tag{3.28}$$

- Conservation of material momentum (for translations in material space):

$$\frac{\partial \mathbf{P}}{\partial t}\bigg|_X - div_R \,\mathbf{b} = \mathbf{f}^{\text{inh}} + \mathbf{f}^{\text{ext}},\tag{3.29}$$

where the have defined the following quantities:

$$H = K^R + W, \quad K^R = \frac{1}{2} \rho_0 \, \mathbf{v}^2, \quad Q = \mathbf{T} \cdot \mathbf{v}, \quad f^{\text{ext}} = -\rho_0 \, \mathbf{f} \cdot \mathbf{F}, \quad (3.30)$$

$$\mathbf{P} = -\rho_0 \, \mathbf{v} \cdot \mathbf{F}, \quad \mathbf{b} = -(L \, \mathbf{1}_R + \mathbf{T} \cdot \mathbf{F}), \quad (3.31)$$

and

$$h = -\left(\frac{K^R}{\rho_0}\right) \left.\frac{\partial \rho_0}{\partial t}\right|_X + \left.\frac{\partial \overline{W}}{\partial t}\right|_{\text{expl}} + \rho_0 \, \mathbf{f} \cdot \mathbf{v}, \quad (3.32)$$

$$\mathbf{f}^{\text{inh}} = \left(\frac{K^R}{\rho_0}\right) \left.\frac{\partial \rho_0}{\partial \mathbf{X}}\right|_t - \left.\frac{\partial \overline{W}}{\partial \mathbf{X}}\right|_{\text{expl}}. \quad (3.33)$$

In view of (3.29) we may say that material momentum is not *strictly* conserved for a materially inhomogeneous elastic material and in the presence of an external body force. It might not be strictly conserved in the presence of distributed or localized material inhomogeneities, but with zero external forces. As to (3.28), it tells us that energy is not strictly conserved for rheonomic materials, whether they suffer growth or ageing. No growth phenomenon is considered in the present book. As to rheonomic materials, the case of so-called *dynamic materials* will be considered, but exceptionally, in the sequel.

Remark 3.5. In the present case, Noether's identity (for the material-space parametrization part of the theorem) is none other than the equation

$$\left(\frac{\partial \mathbf{p}_R}{\partial t} - div_R \, \mathbf{T} - \rho_0 \, \mathbf{f}\right) \cdot \mathbf{F} + \left(\frac{\partial \mathbf{P}}{\partial t} - div_R \, \mathbf{b} - \mathbf{f}^{\text{inh}} - \mathbf{f}^{\text{ext}}\right) \equiv \mathbf{0}. \quad (3.34)$$

Of course, fulfilment of $(3.24)_1$ yields (3.29) for any nonvanishing $\mathbf{F}$. First we note that (3.34) is none other than a reproduction of the manipulation effected in Section 3.2. Second, a very restricted version of (3.34) was given by Ericksen (1977) (quasi-statics, no external force and no material inhomogeneity) as (in our notation)

$$(div_R \, \mathbf{T}) \cdot \mathbf{F} + div_R \, \mathbf{b} = 0. \quad (3.35)$$

But much more than that, Equation (3.34) tells us what to do in the case where we have no variational principle at our disposal (dissipative cases, e.g., in presence of viscosity) in order to mimic Noether's identity and still obtain a (then non-) conservation of pseudomomentum on the material manifold.

3.4 The case of small strains

Here the reference configuration is assimilated to a Lagrangian configuration in physical space. The placement $\mathbf{x}$ remains in a neighbourhood of $\mathbf{x}_0$, the displacement $\mathbf{u} = \mathbf{x} - \mathbf{x}_0$ is infinitesimally small (compared to a macroscopic length), and the strains themselves $\mathbf{e} = (\nabla \mathbf{u})_S$ are small. We need no longer distinguish between lower and upper case Latin tensor indices. The local balance of physical linear momentum and material (pseudo) momentum will now read in Cartesian tensor components as

$$\frac{\partial p_i}{\partial t} - \frac{\partial \sigma_{ji}}{\partial x_j} = \rho_0 \, f_i, \tag{3.36}$$

and

$$\frac{\partial P_i}{\partial t} - \frac{\partial b_{ji}}{\partial x_j} = f_i^{\text{inh}} + f_i^{\text{ext}}, \tag{3.37}$$

with

$$p_i = \rho_0 \, \dot{u}_i, \qquad P_i = -p_i \, F_{ij}, \tag{3.38}$$

and

$$b_{ji} = -(L \, \delta_{ji} + \sigma_{jk} \, F_{ki}), \qquad f_i^{\text{ext}} = -\rho_0 \, f_j \, F_{ji}, \tag{3.39}$$

where

$$F_{ij} = \delta_{ij} + u_{i,j}. \tag{3.40}$$

The second of (3.38) can be written as

$$P_i = -p_i + p_i^w, \qquad p_i^w := -\rho_0 \, \dot{u}_k \, u_{k,i}. \tag{3.41}$$

The last one of these defines the *wave momentum* density. Because (3.36) is satisfied, this momentum $\mathbf{p}^w$ satisfies the reduced equation

$$\frac{\partial}{\partial t} p_i^w - \frac{\partial}{\partial x_j} b_{ji}^w = f_i^{\text{inh}} + f_i^w, \tag{3.42}$$

wherein

$$b_{ji}^w := -(L \, \delta_{ji} + \sigma_{jk} \, u_{k,i}), \qquad f_i^w = -\rho_0 \, f_j \, u_{j,i}. \tag{3.43}$$

It is recalled that

$$L = \frac{1}{2} \rho_0 \, \dot{\mathbf{u}}^2 - W, \qquad W = \overline{W}(\mathbf{e}; \mathbf{x}, t). \tag{3.44}$$

For a medium that is neither inhomogeneous nor rheonomic, and is free of external forces, both (3.37) and (3.42) reduce to strict conservation laws (in the mathematical sense, i.e., with vanishing right-hand side):

$$\frac{\partial p_i}{\partial t} - \frac{\partial \sigma_{ji}}{\partial x_j} = 0, \qquad \frac{\partial p_i^w}{\partial t} - \frac{\partial b_{ji}}{\partial x_j} = 0, \tag{3.45}$$

where the second is deduced from the first by left multiplication by $u_{i,k}$ and renaming the indices (remember Noether's identity). Still each of these equations is used for a different purpose, the first one to study solutions of a problem with prescribed boundary and initial conditions, and the second one in a post-processing strategy. Note that in linear elasticity the first of (3.45) is linear in the space-time derivatives of the displacement, while the second of (3.45) is jointly *quadratic* in these derivatives. Remark that we simply note b_{ji}^w as b_{ji} without risk of ambiguity and continue to call it Eshelby stress although any reference to a material frame has disappeared.

3.5   Peculiarity of a one-dimensional motion

Consider a problem in which all fields are only functions of one space coordinate (say x) and time. The displacement **u** may be reduced to a scalar component u but it is not necessarily along the direction x. For example, we may have to examine an elastic wave motion which is related to a transverse elastic displacement only (e.g., SH surface waves in geology). We note u_x and u_t the space and time derivatives of u. Also

$$\sigma = \frac{\partial W}{\partial e} = E\,u_x, \quad W = \frac{1}{2}\,\overline{E}(\mathbf{x},t)\,(u_x)^2, \quad K = \frac{1}{2}\,\bar{\rho}_0(\mathbf{x},t)\,(u_t)^2, \quad (3.46)$$

where E is the only surviving coefficient of elasticity. We have

$$L = \frac{1}{2}\,\bar{\rho}_0(x,t)\,(u_t)^2 - \frac{1}{2}\,\overline{E}(x,t)\,(u_x)^2, \quad (3.47)$$

and

$$H = \frac{1}{2}\,\bar{\rho}_0(x,t)\,(u_t)^2 + \frac{1}{2}\,\overline{E}(x,t)\,(u_x)^2. \quad (3.48)$$

In the absence of external force the linear (physical) momentum equation obviously yields

$$\frac{\partial}{\partial t}\left(\bar{\rho}_0(x,t)\,\frac{\partial u}{\partial t}\right) - \frac{\partial}{\partial x}\left(\overline{E}(x,t)\,\frac{\partial u}{\partial x}\right) = 0 \quad (3.49)$$

while the equations of "conservation" of energy and wave momentum reduce to

$$\frac{\partial H}{\partial t} - \frac{\partial Q}{\partial x} = h := \left.\frac{\partial L}{\partial t}\right|_{\text{expl}} \quad (3.50)$$

and

$$\frac{\partial p^w}{\partial t} - \frac{\partial b}{\partial x} = f := \left.\frac{\partial L}{\partial x}\right|_{\text{expl}} \quad (3.51)$$

with canonical quantities $\mathbf{Q}$, $\mathbf{p}^w$, and $\mathbf{b}$ reduced to scalars such that ($cf.$ Appendix A for canonical definitions)

$$Q = -u_t \frac{\partial L}{\partial u_x} = \overline{E}\, u_t\, u_x, \tag{3.52}$$

$$P = -u_x \frac{\partial L}{\partial u_t} = -\bar{\rho}_0\, u_t\, u_x, \tag{3.53}$$

$$b = -\left(L - u_x \frac{\partial L}{\partial u_x}\right) = -(L + \sigma\, u_x). \tag{3.54}$$

Setting $c_0^2 = \overline{E}/\bar{\rho}_0$, we note that in this "scalar" case

$$Q = -c_0^2\, p^w, \qquad b = -H, \tag{3.55}$$

so that in the $scleronomic\ homogeneous$ case equations (3.50) and (3.51) reduce to the following remarkable system:

$$\frac{\partial H}{\partial t} + c_0^2 \frac{\partial p^w}{\partial x} = 0, \qquad \frac{\partial p^w}{\partial t} + \frac{\partial H}{\partial x} = 0. \tag{3.56}$$

Furthermore, while the displacement $\mathbf{u}$ satisfies the standard linear wave equation

$$u_{tt} - c_0^2\, u_{xx} = 0, \tag{3.57}$$

this is also the case of the two scalars H and p^w, that is,

$$H_{tt} - c_0^2\, H_{xx} = 0, \qquad (p^w)_{tt} - c_0^2\, (p^w)_{xx} = 0, \tag{3.58}$$

as shown by elimination between the two equations (3.56); in other words, both quantities that are linear and quadratic in the derivatives of u are transported in the same wavelike manner. This remarkable result and the symmetry present in (3.56) are however misleading because we are in a special case where both energy and wave-momentum equations reduce to scalar equations, whereas in a general framework they are a scalar equation and a co-vectorial equation, respectively. We owe to one of our teachers, W. D. Hayes (1974a, pp. 23–24) the direct introduction of the couple of equations (3.56) which he suggested then as two $quadratic\text{-}invariant$ equations deduced from (3.57) — without the present "Eshelbian" framework and the consequence in (3.58). The more general equations (3.50) and (3.51) will be useful when dealing with "dynamic" materials.

3.6 Small strains in the presence of dissipation

For the sake of illustration we consider the case of elastic bodies endowed with some viscosity, hence visco-elastic bodies, but in small strains. Imagine that the Cauchy stress is given in additive form by

$$\sigma_{ji} = \sigma_{ji}^{\text{elas}} + \sigma_{ji}^{\text{visc}}, \tag{3.59}$$

where the elastic contribution may be given by an expression such as (2.96) while the dissipative part of the Kelvin–Voigt type (for this, see e.g., Eringen (1980, Chapter 9)) is given for isotropic bodies by

$$\sigma_{ji}^{\text{visc}} = \lambda_\nu \, \dot{e}_{kk} \, \delta_{ij} + 2 \, \mu_\nu \, \dot{e}_{ij}. \tag{3.60}$$

Here λ_ν and μ_ν are two viscosity coefficients that, upon account of the reduced form of the inequality (2.59), i.e.,

$$\Phi^{\text{visc}} = \sigma_{ij}^{\text{visc}} \, \dot{e}_{ji} \geq 0, \tag{3.61}$$

must satisfy the inequalities (compare to (2.97))

$$3\,\lambda_\nu + 2\,\mu_\nu \geq 0, \qquad \mu_\nu \geq 0. \tag{3.62}$$

Equation (3.36) still holds true but with (3.59) valid. The system is dissipative and (3.36) cannot be obtained by means of a standard variational principle so that Noether's theorem does not apply. To obtain a sensible equation of conservation (or non-conservation) of wave momentum, we can only mimic the "prescription" typically represented by equation (3.34) but in small strains. We consider the case of a homogeneous scleronomic body for the sake of simplicity. That is, we apply $u_{i,k}$ to all terms in Equation (3.36), sum over dummy indices, and note the following intermediate results:

$$\frac{\partial p_i}{\partial t} \, u_{i,k} = -\frac{\partial p_k^w}{\partial t} - \frac{\partial}{\partial x_k} \left(\frac{1}{2} \, \rho_0 \, \dot{\mathbf{u}}^2 \right),$$

$$\frac{\partial \sigma_{ji}}{\partial x_j} \, u_{i,k} = \frac{\partial}{\partial x_j} (\sigma_{ji} \, u_{i,k}) - \sigma_{ji}^{\text{elas}} \, u_{i,kj} - \sigma_{ji}^{\text{visc}} \, u_{i,kj}$$

$$= \frac{\partial}{\partial x_j} (\sigma_{ji} \, u_{i,k}) - \frac{\partial W^{\text{elas}}}{\partial e_{ji}} (u_{i,j})_{,k} - \sigma_{ji}^{\text{visc}} \, u_{i,jk},$$

$$\frac{\partial W^{\text{elas}}}{\partial e_{ji}} (u_{i,j})_{,k} = \frac{\partial}{\partial x_j} (W \, \delta_{jk}), \qquad \sigma_{ji}^{\text{elas}} = \frac{\partial W^{\text{elas}}}{\partial e_{ij}}.$$

Accounting for these and rearranging the indices, we obtain the looked for (non-conservation) equation of wave momentum in the form

$$\frac{\partial}{\partial t}p_i^w - \frac{\partial}{\partial x_j}b_{ji}^w = f_i^{w.b} + f_i^{w.\text{dissip}}, \qquad (3.63)$$

wherein

$$b_{ji}^w := -(L^{\text{elas}}\,\delta_{ji} + \sigma_{jk}\,u_{k,i}),$$
$$f_i^{w.b} = -\rho_0\,f_j\,u_{j,i}, \qquad f_i^{w.\text{dissip}} := \sigma_{kj}^{\text{visc}}\,u_{j,ki}. \qquad (3.64)$$

It is recalled that

$$L^{\text{elas}} = \frac{1}{2}\rho_0\,\dot{\mathbf{u}}^2 - W^{\text{elas}}, \qquad W^{\text{elas}} = \overline{W}(\mathbf{e}), \qquad \sigma_{ji}^{\text{elas}} = \frac{\partial \overline{W}}{\partial e_{ij}}. \qquad (3.65)$$

Here the effect of viscosity is interpreted as a *pseudo-material inhomogeneity* since it appears as a source "material" force in the right-hand side of Equation (3.63).

A parallel manipulation effected on equation (3.36) by taking its inner product with the velocity and accounting for (3.65) and the partial result

$$\frac{\partial \sigma_{ji}}{\partial x_j}\,\dot{u}_i = \frac{\partial}{\partial x_j}(\sigma_{ji}\,\dot{u}_i) - \frac{\partial W^{\text{elas}}}{\partial t} - \sigma_{ji}^{\text{visc}}\,\dot{u}_{i,j}$$

leads to the following completely compatible — with (3.63) — expression

$$\frac{\partial}{\partial t}\left(\frac{1}{2}\rho_0\,\dot{\mathbf{u}}^2 + W^{\text{elas}}\right) - \frac{\partial}{\partial x_j}(\sigma_{ji}\,\dot{u}_i) = -\sigma_{ji}^{\text{visc}}\,\dot{e}_{ij} + \rho_0\,f_i\,\dot{u}_i \qquad (3.66)$$

where, obviously, viscous dissipation is none other than a source in the right-hand side of this equation. Note again that it is the *total* (elastic plus viscous) stress that is involved in the divergence term of the left-hand side of (3.66) just like in the expression of b_{ji}^w.

Such a scheme using equations (3.63) and (3.66) has been exploited for elastic surface waves propagating over a visco-elastic substrate (Rousseau and Maugin (2012a); see also Section 8.6 herein after).

Chapter 4

Action, phonons and wave mechanics

Object of the chapter

This chapter is quite original in that it expands notions that are not so much exploited in standard continuum mechanics. These are: the wave-particle dualism, the notion of action and its local conservation, the wave kinematics in the manner of Lighthill and Whitham, the wave action, the material wave momentum and the associated wave-Eshelby stress, the evolution of the wave amplitude, and useful related techniques such as those of the averaged Lagrangian and the two-timing method.

Preliminary remark: Physical dimensions

Let M, L and T denote the usual symbols of dimension for mass, length and time. Then we have:

$$\text{Dim } [k = \text{wave number}] = L^{-1},$$
$$\text{Dim } [\omega = \text{circular frequency}] = T^{-1},$$
$$\text{Dim } [\varphi = \text{phase}] = 1,$$

and

$$\text{Dim } [v = \text{velocity}] = L\,T^{-1},$$
$$\text{Dim } [\gamma = \text{acceleration}] = L\,T^{-2},$$
$$\text{Dim } [F = \text{force}] = M\,L\,T^{-2}.$$

For a point particle:

$$\text{Dim } [p = \text{momentum}] = M\,L\,T^{-1},$$
$$\text{Dim } [E = \text{energy}] = M\,L^2\,T^{-2},$$
$$\text{Dim } [P = \text{power}] = \text{Dim } [\text{energy / time}] = M\,L^2\,T^{-3},$$
$$\text{Dim } [A = \text{action}] = \text{Dim } [\text{energy} \times \text{time}] = M\,L^2\,T^{-1}.$$

In a 3D continuum (quantities per unit volume):

$$\text{Dim [density]} = M\,L^{-3},$$
$$\text{Dim [momentum]} = M\,L^{-2}\,T^{-1},$$
$$\text{Dim [energy]} = M\,L^{-1}\,T^{-2},$$
$$\text{Dim [power]} = M\,L^{-1}\,T^{-2},$$
$$\text{Dim [action]} = M\,L^{-1}\,T^{-1},$$
$$\text{Dim [stress]} = \text{Dim [force per unit surface]} = M\,L^{-1}\,T^{-2}$$
$$= \text{Dim [energy per unit volume]}.$$

In a 3D continuum (for extensive quantities; quantities per unit mass): Divide those above by the density.

4.1 Wave-particle dualism and phonons

Early in the twentieth century Max Planck (1858–1947, Nobel Prize 1918) introduced the quantum hypothesis in the theory of radiation through the formula

$$E = h\,\nu = \hbar\omega, \tag{4.1}$$

where E is the energy, ω is the circular frequency, and the universal constant $\hbar = h/2\,\pi$ is called the reduced elementary (quantum of) action (see above the physical dimension). This notion was soon applied by Einstein (1905) in his theory of the photo-electric effect according to which light energy is radiated in quanta, i.e., in definite bursts. This interpretation associates the notion of "grain" with that of a wave: photons are grains of light; the photo-electric effect takes place through the emission of "grains of light", a view that would have pleased Newton. With a remarkable insight Louis de Broglie (1892–1987, NP 1929; see his thesis — de Broglie (1924) — that followed a series of Notes) introduced the idea that a momentum p could be associated with such a wave-like phenomenon *via* the relation

$$p = h/\lambda = \hbar\,k, \tag{4.2}$$

where λ is the wavelength and k is the wave number associated with the wave. But the real breakthrough here is the fact that de Broglie proposed that both (4.1) and (4.2) hold good for *any* "particle" or wavelike phenomenon, light or other. The main formula also is that since $\hbar$ is common to both formulas, the *wave phase*

$$\varphi = k\,x - \omega\,t, \tag{4.3}$$

and the *action* A are such that

$$A = p\,x - E\,t = \hbar\,\varphi. \tag{4.4}$$

That is, they are proportional to one another. The *wave-particle dualism* thus introduced opened the path for the construction of *wave mechanics* (in French: *mécanique ondulatoire*) mostly at the hands of Erwin Schrödinger (1887–1961, NP 1933), while being experimentally confirmed by C. J. Davisson (NP 1937) and L. H. Germer in their successful experiment of the diffraction of electrons by a crystal. Note that both φ and A are scalar Lorentz invariant, a fact that delighted Einstein when Paul Langevin communicated to him de Broglie's thesis.

Igor Tamm (1922–1995, NP 1958) — a Russian scientist — introduced the notion of **phonons**: phonons are quasi-particles representing the quantization of the normal modes of lattice vibrations of periodic elastic crystalline structures in solids. They are thus related to bulk dynamical elasticity or *acoustic waves in solids*. By analogy with photons, they may be called "grains of acoustic waves" (*cf.* Kittel (2005); Kosevich (1999); Maynard (2000)): while normal modes are wave-like phenomena in classical continuum mechanics, they have particle-like properties in the wave-particle dual description of quantum mechanics. This approach pertains to quantum mechanics, a field of science obviously quite different from traditional continuum mechanics (our main concern).

4.2 Action in continuum mechanics

During all his life de Broglie frequently made use of arguments of analytical mechanics (principle of least action, Maupertuis' ideas, Hamilton–Jacobi equation, *etc.*) of which he was very fond. It is also our case in this book but in continuum mechanics. The essential ingredient of analytical continuum mechanics is the total *Hamiltonian–Lagrangian action* in the form

$$HA = \int_B \int_T L\, dV\, dt, \tag{4.5}$$

where B is a three-dimensional domain of Euclidean space E^3, T is an interval of time, and L is a Lagrangian volume density

$$L = K - W. \tag{4.6}$$

Here K is the volume kinetic energy and W is the volume potential energy. An example of a more specific expression for HA was given in Chapter 3 were the volume B was material. Consequences of this last formulation were therein the local field equations (in the Piola format) and additional conservation equations on the material manifold. Of particular interest are the equations of material momentum and energy, which indeed form globally a four-dimensional conservation law in the proper space-time parametrization (material coordinates and Newtonian time). They govern the material momentum $\mathbf{P}$ and the total energy — or Hamiltonian density — H, both per unit referential volume. We remind the reader that for elasticity they read

$$\frac{\partial}{\partial t}\mathbf{P}\bigg|_{\mathbf{X}} - div_R\,\mathbf{b} = \mathbf{f}^{\text{inh}}, \tag{4.7}$$

and

$$\frac{\partial}{\partial t}H\bigg|_{\mathbf{X}} - \nabla_R \cdot \mathbf{Q} = h, \tag{4.8}$$

wherein

$$\mathbf{P} := -\rho_0\,\mathbf{v}\cdot\mathbf{F}, \qquad \mathbf{b} = -(L\,\mathbf{1}_R + \mathbf{T}\cdot\mathbf{F}), \tag{4.9}$$

$$H = K + W, \qquad \mathbf{Q} = \mathbf{T}\cdot\mathbf{v}, \tag{4.10}$$

$$h := -\frac{\partial L}{\partial t}\bigg|_{\text{expl}} = -\frac{K}{\rho_0}\frac{\partial\hat\rho_0}{\partial t}\bigg|_X + \frac{\partial\widehat{W}}{\partial t}\bigg|_{\text{expl}}, \tag{4.11}$$

$$\mathbf{f}^{\text{inh}} := \frac{\mathbf{v}^2}{2}\nabla_R\bar\rho_0 - \frac{\partial\overline{W}}{\partial\mathbf{X}}\bigg|_{\text{expl}}, \tag{4.12}$$

with a rather general Lagrangian material density

$$L(\mathbf{v},\mathbf{F};\mathbf{X},t) = K(\mathbf{v};\mathbf{X},t) - \widehat{W}(\mathbf{F};\mathbf{X},t), \tag{4.13}$$

with kinetic energy

$$K(\mathbf{v};\mathbf{X},t) = \frac{1}{2}\hat\rho_0(\mathbf{X},t)\,\mathbf{v}^2. \tag{4.14}$$

Elastic systems with Lagrangian (4.13) are said to be *rheonomic* and *materially inhomogeneous* systems because of the explicit dependency on $\mathbf{X}$ and t.

Conservation of action

In continuum mechanics the density of canonical action per unit referential volume will be defined by

$$A := \mathbf{P} \cdot \mathbf{X} - H\, t, \tag{4.15}$$

while the *phase* for plane travelling waves is usually defined by

$$\varphi = \mathbf{K} \cdot \mathbf{X} - \omega\, t, \tag{4.16}$$

where $\mathbf{K}$ now is a material wave vector and ω is a circular frequency; then the analogy between (4.4) and (4.3) is crystal-clear. From this analogy we infer that the action (4.15) should play a prevailing role in wave studies in dynamic (bulky) materials, although this is very seldom considered by mechanicians and engineers. Indeed, on taking the scalar product of (4.7) by $\mathbf{X}$ in material space, and the product of (4.8) by t and subtracting the latter result from the former and performing a few manipulations, we establish the following (non-strict) **conservation law** for the continuum action:

$$\left.\frac{\partial}{\partial t}A\right|_{\mathbf{X}} - \nabla_R \cdot (\mathbf{b} \cdot \mathbf{X} - \mathbf{Q}\,t) = \mathbf{f}^{\text{inh}} \cdot \mathbf{X} - h\,t - (H + tr\,\mathbf{b}). \tag{4.17}$$

In the proof of this result we have accounted for the fact that $\mathbf{X}$ and t are independent of one another in the present space-time parametrization $(\mathbf{X}, t)$. Equation (4.17) becomes a *strict* conservation law (no source in the right-hand side) in the following circumstances. First, for a *scleronomic* (contrary of rheonomic) and homogeneous system, the first two terms in the right-hand side of this equation vanish. Second, the last term within parentheses vanishes identically in the case of one space dimension — for which $tr\,\mathbf{b} = b = -H$ as noted in Chapter 3 — see below. Otherwise (4.17) never is a strict conservation law. Furthermore, even *in quasi statics*, this equation with nonzero right-hand side plays a role in some mathematical proof. Indeed, the resulting identity for homogeneous materials then reads

$$\nabla_R \cdot (\mathbf{b} \cdot \mathbf{X}) - (W + tr\,\mathbf{b}) = 0, \qquad \mathbf{b} = W\,\mathbf{1}_R - \mathbf{T} \cdot \mathbf{F}. \tag{4.18}$$

The integrated form of this over a material body is of interest in some mathematical proofs — see Knops *et al.* (2003).

Consider the *one-dimensional case in small strains*. Accordingly, we work in two-dimensional Euclidean space-time (x, t). With a standard notation (u = elastic displacement; derivatives indicated by a subscript x or t), the Lagrangian density (4.13) reduces to

$$L = \frac{1}{2}\,\hat{\rho}_0(x,t)\,(u_t)^2 - \frac{1}{2}\,\widehat{E}(x,t)\,(u_x)^2, \tag{4.19}$$

and the canonical quantities $\mathbf{Q}$, $\mathbf{P}$, and $\mathbf{b}$ become scalars such that

$$H = \frac{1}{2}\,\hat{\rho}_0(x,t)\,(u_t)^2 + \frac{1}{2}\,\widehat{E}(x,t)\,(u_x)^2, \tag{4.20}$$

$$Q = -u_t\,\frac{\partial L}{\partial u_x} = \widehat{E}\,u_t\,u_x = \sigma\,u_t, \tag{4.21}$$

$$P = -u_x\,\frac{\partial L}{\partial u_t} = -\hat{\rho}_0\,u_t\,u_x = -p\,u_x, \tag{4.22}$$

$$b = -\left(L - u_x\,\frac{\partial L}{\partial u_x}\right) = -H, \tag{4.23}$$

while the one-dimensional wave equation reads

$$\frac{\partial}{\partial t}\left(\hat{\rho}_0(x,t)\,\frac{\partial u}{\partial t}\right) - \frac{\partial}{\partial x}\left(\widehat{E}(x,t)\,\frac{\partial u}{\partial x}\right) = 0. \tag{4.24}$$

For a *scleronomic materially homogeneous* system, Equation (4.17) now reduces to the strict conservation law of action:

$$\left.\frac{\partial}{\partial t}A\right|_X - \frac{\partial}{\partial x}(b\,x - Q\,t) = 0, \qquad A := P\,x - H\,t. \tag{4.25}$$

Integrating the first of these over the real line, and for a field solution tending sufficiently fast to zero at infinities, we obtain a *global conservation of action* in the form

$$\frac{d}{dt}\int_R A\,dX = 0. \tag{4.26}$$

4.3   Wave kinematics and wave action

The Hamiltonian action HA on which the formulation (4.7)–(4.14) is based reads

$$HA = \int_t\int_{B_R} L\left(\frac{\partial\overline{\mathbf{x}}}{\partial\mathbf{X}} = \mathbf{F},\,\frac{\partial\overline{\mathbf{x}}}{\partial t} = \mathbf{v};\mathbf{X},t\right) dV\,dt. \tag{4.27}$$

Let us consider Whitham's (1965; 1974) theory of the so-called *averaged Lagrangian* and the corresponding wave system. It is assumed there that wave parameters such as frequency, wave number and amplitude vary slowly in space and time in the sense that their relative changes in one wavelength and in one period are small. We are dealing with slowly varying wave trains with one phase function φ. Accordingly, derivatives of the frequency, wave number and amplitude are small in an appropriate mathematical sense and can thus be neglected. A two-time perturbation scheme presented by Whitham justifies this approximation (see Appendix B). In this *kinematic wave theory* due mostly to Lighthill (1965); Whitham (1965, 1974), and Hayes (1970a,b, 1973, 1974b) a general phase function

$$\varphi = \overline{\varphi}(\mathbf{X}, t), \tag{4.28}$$

is introduced from which the material wave vector $\mathbf{K}$ and the frequency ω are defined by

$$\mathbf{K} = \frac{\partial \overline{\varphi}}{\partial \mathbf{X}} = \nabla_R \overline{\varphi}, \qquad \omega = -\frac{\partial \overline{\varphi}}{\partial t}. \tag{4.29}$$

Whence there follows at once the two equations

$$\nabla_R \times \mathbf{K} = \mathbf{0}, \tag{4.30}$$

and

$$\frac{\partial \mathbf{K}}{\partial t} + \nabla_R \omega = \mathbf{0}. \tag{4.31}$$

The first of these reflects the curl-free nature of $\mathbf{K}$, while the second is in the form of a strict conservation that we may call the *conservation of wave vector*.

Of course, Equations (4.29) are trivially satisfied for plane wave solutions with an *a priori* phase $\overline{\varphi} = \mathbf{K} \cdot \mathbf{X} - \omega t$.

For an *inhomogeneous rheonomic* linear behavior with dispersion we would have a dispersion relation $\omega = \Omega(\mathbf{K}; \mathbf{X}, t)$ while in an inhomogeneous rheonomic dispersive **nonlinear** material, the frequency will also depend on the amplitude. Let $\mathbf{a}$ the n-vector of $\mathbb{R}^n$ that characterizes this small slowly varying amplitude of a complex system (in general with several degrees of freedom). Thus, now,

$$\omega = \Omega(\mathbf{K}; \mathbf{X}, t; \mathbf{a}). \tag{4.32}$$

For a wave motion depending on the phase (4.28) and with all characteristic quantities varying slowly over space-time, Whitham proposes to replace the initial variational problem, say that based on a Hamiltonian–Lagrangian action (4.27) for elasticity, by one pertaining to the averaged Lagrangian, i.e.,

$$\delta \int \widetilde{L} \, d\mathbf{X} \, dt = 0, \qquad \widetilde{L} = \langle L \rangle := \frac{1}{2\pi} \int_0^{2\pi} L \, d\varphi, \tag{4.33}$$

with (compare to the integrand in (4.27))

$$\widetilde{L} = \widetilde{L}\left(\frac{\partial \overline{\varphi}}{\partial \mathbf{X}} = \mathbf{K}, \frac{\partial \overline{\varphi}}{\partial t} = -\omega, \mathbf{a}; \mathbf{X}, t\right), \tag{4.34}$$

where now the *fields* are the *amplitude* **a** and the *phase* φ. Accordingly, the Euler–Lagrange equations based on this Lagrangian read as the following *field equations*:

$$\frac{\partial \widetilde{L}}{\partial \mathbf{a}} = \mathbf{0}, \tag{4.35}$$

$$\frac{\partial \widetilde{S}}{\partial t} - \nabla_R \cdot \mathbf{W} = 0, \tag{4.36}$$

obtained by direct variation. Here we have set

$$\widetilde{S} := \frac{\partial \widetilde{L}}{\partial \omega}, \qquad \mathbf{W} := \frac{\partial \widetilde{L}}{\partial \mathbf{K}}. \tag{4.37}$$

On applying Noether's theorem for time translations and material-space translations we are led to the fooling two (non-strict) *conservation laws* of energy and canonical momentum (*cf.* Maugin (2007)):

$$\frac{\partial \widetilde{H}}{\partial t} - \nabla_R \cdot \widetilde{\mathbf{Q}} = \tilde{h}, \qquad \frac{\partial \widetilde{\mathbf{P}}}{\partial t} - div_R \, \widetilde{\mathbf{b}} = \widetilde{\mathbf{f}}^{\text{inh}}, \tag{4.38}$$

wherein

$$\widetilde{H} = \omega \, \widetilde{S} - \widetilde{L}, \quad \widetilde{\mathbf{Q}} = \omega \, \mathbf{W}, \quad \tilde{h} = -\frac{\partial \widetilde{L}}{\partial t}\bigg|_{\text{expl}}, \tag{4.39}$$

$$\widetilde{\mathbf{P}} = \widetilde{S} \, \mathbf{K}, \quad \widetilde{\mathbf{b}} = -\left(\widetilde{L} \, \mathbf{1}_R - \mathbf{W} \otimes \mathbf{K}\right), \quad \widetilde{\mathbf{f}}^{\text{inh}} = \frac{\partial \widetilde{L}}{\partial \mathbf{X}}\bigg|_{\text{expl}}. \tag{4.40}$$

Dimensionally, $\widetilde{S}$ is an *action* and may be called the **wave action**, while Equation (4.36) may be referred to as a strict *conservation law* for the wave action in which **W** is the **action flux**. The material co-vector $\widetilde{\mathbf{P}}$ may be called the **material wave momentum** (notice that its formula reminds us of the already mentioned quantum wave-mechanics relationship due to de Broglie: $\widetilde{\mathbf{P}} = \hbar \, \mathbf{K}$), and $\widetilde{\mathbf{b}}$, the associated flux, may be called the **material wave Eshelby stress**. This tensor is not symmetric unless **W** is proportional to **K**. The wave action conservation equation (4.36) plays here the central role (equivalent to the balance of linear physical momentum in standard continuum mechanics). Indeed, in the same way as Eq. (3.28) and (3.29) for elasticity can be deduced from the balance of physical linear momentum (3.24) by right scalar multiplication by **v** and **F**, Equations (4.38)$_{1-2}$ can be deduced from (4.36) by simple scalar and tensorial multiplication, respectively, by ω and **K** on account of the functional dependency assumed for $\widetilde{L}$ (*cf.* Maugin (2007, 2008)). For instance, we can note the following "Noether's identity":

$$\left\{\frac{\partial \widetilde{S}}{\partial t} - (\nabla_R \cdot \mathbf{W})\right\} \mathbf{K} - \left\{\frac{\partial \widetilde{\mathbf{P}}}{\partial t} - div_R \, \widetilde{\mathbf{b}} - \frac{\partial \widetilde{L}}{\partial \mathbf{X}}\bigg|_{\text{expl}}\right\} \equiv \mathbf{0}. \tag{4.41}$$

Of course, reasoning in $(\mathbf{K}, \omega)$ space is somewhat dual to that in space-time $(\mathbf{X}, t)$, the duality being understood in the sense of the phase function, hence of the action function according to the kinematic-wave theory. A final remark is in order: taking the inner product with $\mathbf{X}$ in material space of the second of (4.40) and combining with the product of the first of (4.39) by time t, we obtain the following remarkable result:

$$A = \widetilde{\mathbf{P}} \cdot \mathbf{X} - \widetilde{H} t = \widetilde{S} (\mathbf{K} \cdot \mathbf{X} - \omega t) + \widetilde{L} t. \tag{4.42}$$

For a *linear* elastic body in 1D (case of phonons) with $u = \mathbf{a} \cos \varphi$, the Lagrangian density is quadratic in the derivatives of u. It is immediately shown that (Maugin (2007, Eqs. (4.5)–(4.6))) here K is the wave number no to be confused with the kinetic energy and we keep a bold $\mathbf{a}$ for the amplitude although it is a scalar)

$$\widetilde{L} = D_L(\omega, K) \, \mathbf{a}^2, \tag{4.43}$$

where $D_L(\omega, K) = 0$ happens to be the "linear" dispersion relation [as a matter of fact, the "field equation" associated with the amplitude; *cf.* Eq. (4.35). Accordingly, in this case (4.42) reduces to

$$A = \widetilde{\mathbf{P}} \cdot \mathbf{X} - \widetilde{H} t = \widetilde{S} (\mathbf{K} \cdot \mathbf{X} - \omega t) = \widetilde{S} \varphi, \tag{4.44}$$

in agreement with the Planck–de Broglie formula of proportionality between action and phase — *cf.* Equation (4.4).

The more general formula (4.42) is still admissible as the right-hand side may be rewritten as (compare Eq. (4.3) in Landau and Lifshitz (1965))

$$A = \widetilde{\mathbf{P}} \cdot \mathbf{X} - \widetilde{H} t = \widetilde{S} \mathbf{K} \cdot \mathbf{X} - \widetilde{H} t, \tag{4.45}$$

where the first contribution in the right-hand side may be considered to be the "reduced" action density of Landau and Lifshitz for the present formulation. Such notions can prove useful in some wave problems.

4.4 Evolution equation for the wave amplitude

In a general linear *dispersive* 1D case with a dispersion equation written as

$$D_L(\omega, K) = 0, \quad \text{or } \omega = \Omega(K), \tag{4.46}$$

the *group velocity* is given by

$$\nu_g = \frac{\partial \Omega}{\partial K} = -\frac{\partial D_L / \partial K}{\partial D_L / \partial \omega}. \tag{4.47}$$

Thus, according to (4.43),

$$\frac{\partial \widetilde{L}}{\partial \omega} = \frac{\partial D_L}{\partial \omega} \mathbf{a}^2, \qquad \frac{\partial \widetilde{L}}{\partial K} = \frac{\partial D_L}{\partial K} \mathbf{a}^2. \tag{4.48}$$

It follows from this that (4.36) yields

$$\frac{\partial}{\partial t} \left(\frac{\partial D_L}{\partial \omega} \mathbf{a}^2 \right) - \frac{\partial}{\partial x} \left(\frac{\partial D_L}{\partial K} \mathbf{a}^2 \right) = 0. \tag{4.49}$$

Because of (4.47) and (4.31), this is readily transformed into the following (strict) *conservation equation* for the square of the amplitude:

$$\frac{\partial \mathbf{a}^2}{\partial t} + \frac{\partial}{\partial x} (\nu_g(K) \mathbf{a}^2) = 0. \tag{4.50}$$

Because energy density is proportional to $\mathbf{a}^2$, this is tantamount to saying that energy propagates with the group velocity. By the same token (4.21) yields

$$\frac{DK}{Dt} = 0; \qquad \frac{D}{Dt} := \frac{\partial}{\partial t} + \nu_g \frac{\partial}{\partial x}. \tag{4.51}$$

This means that K is constant along characteristic curves defined by $dx/dt = \nu_g$ in the (x, t) plane. As K is constant on each of these curves, these curves indeed are straight lines, each with a slope ν_g. But in the nonlinear case with a dispersion relation now dependent on the amplitude $\mathbf{a}$, i.e. $D(\omega, K, \mathbf{a}) = 0$, Equations (4.50) and (4.21) become coupled. Whitham's variational method also deals with this more general case.

Still in the linear case, the stationary value is $\widetilde{L} = 0$, and the energy density given by the first of (4.39) reduces to $\widetilde{H} = \omega \widetilde{S}$. As a consequence, the wave-action conservation (4.36) can be written as

$$\frac{\partial}{\partial t} \left(\frac{\widetilde{H}}{\omega} \right) + \frac{\partial}{\partial x} \left(\nu_g(K) \frac{\widetilde{H}}{\omega} \right) = 0. \tag{4.52}$$

This is a conservation equation for the so-called *adiabatic invariant* $\widetilde{H}/\omega$ that appears in the study of slow modulations of a linear vibrating system.

4.5 Hamiltonian formulation

We return to the 3D case for an *inhomogeneous rheonomic* linear behavior with dispersion relation in the general function form

$$\omega = \Omega(\mathbf{K}; \mathbf{X}, t). \tag{4.53}$$

The *conservation of wave vector* becomes

$$\frac{\partial \mathbf{K}}{\partial t} + \mathbf{V}_g \cdot \nabla_R \mathbf{K} = -\frac{\partial \Omega}{\partial \mathbf{X}}\bigg|_{\text{expl}}, \qquad \mathbf{V}_g = \frac{\partial \Omega}{\partial \mathbf{K}}, \tag{4.54}$$

from which there follows the Hamiltonian system

$$\frac{D\mathbf{X}}{Dt} = \frac{\partial \Omega}{\partial \mathbf{K}}, \qquad \frac{D\mathbf{K}}{Dt} = -\frac{\partial \Omega}{\partial \mathbf{X}}\bigg|_{\text{expl}}, \tag{4.55}$$

wherein [compare to the second of (4.51)]

$$\frac{D}{Dt} \equiv \frac{\partial}{\partial t} + \mathbf{V}_g \cdot \nabla_R. \tag{4.56}$$

Simultaneously, from the second of (4.29) we have the *Hamilton–Jacobi equation*

$$\frac{\partial \varphi}{\partial t} + \Omega\left(\mathbf{X}, t; \mathbf{K} = \frac{\partial \varphi}{\partial \mathbf{X}}\right) = 0. \tag{4.57}$$

If we now consider a wave in an inhomogeneous rheonomic dispersive *nonlinear* material, the frequency will also depend on the amplitude. Let $\mathbf{a}$ the n-vector of $\mathbf{R}^n$ that characterizes this small slowly varying amplitude of a complex system (in general with several degrees of freedom). Thus, now,

$$\omega = \Omega(\mathbf{K}; \mathbf{X}, t, \mathbf{a}). \tag{4.58}$$

Accordingly, the second of Hamilton's equations (4.55) will now read

$$\frac{D\mathbf{K}}{Dt} = -\frac{\partial \Omega}{\partial \mathbf{X}}\bigg|_{\text{expl}} + \mathbf{A} \cdot \left(\nabla_R \mathbf{a}\right)^T, \qquad \mathbf{A} := -\frac{\partial \Omega}{\partial \mathbf{a}}. \tag{4.59}$$

In the studies of Newell (1985) and Maugin and Hadouaj (1991), one is even led to considering a *nonlinear "dispersive" dispersion relation* in which the assumed slowly varying quantities such as space and time derivative of the amplitude are involved in the function Ω, which relation becomes a true "wave equation" itself for the amplitude — an example of this possibility is given in Chapter 10 herein after.

4.6 Further analytical mechanics

Forgetting about the expressions given in (4.39)–(4.40), consider a general smooth function $S = \overline{S}(\mathbf{X}, t)$ and define a general material momentum and the energy density by [compare to (4.29)]

$$\mathbf{P} = \nabla_R \overline{S}, \qquad H = -\frac{\partial \overline{S}}{\partial t}\bigg|_X. \tag{4.60}$$

where we recall that ∇_R is the material gradient and $d/dt := \partial/\partial t|_X$ is the material time derivative. Obviously then,

$$\nabla_R \times \mathbf{P} = \mathbf{0}, \qquad \frac{d\mathbf{P}}{dt} = -\nabla_R H. \tag{4.61}$$

From the second of these it follows that if the first is valid initially, it remains valid in time. In standard analytical mechanics, the first of (4.60) is none other than the *Jacobi equation of motion*; it would be de Broglie's *"guidance" equation* in the causal interpretation of quantum mechanics (see Holland (1993, Chapter 2); Jammer (1974, Sections 2.5 and 2.6)).

In nonlinear (of course conservative) dynamic inhomogeneous (but scleronomic = no explicit time dependence) elasticity we know the expression of H, for instance as a sufficiently regular function

$$H = \overline{H}(\mathbf{P}, \mathbf{X}, \nabla\mathbf{X}, N), \tag{4.62}$$

where $\nabla\mathbf{X} = \mathbf{F}^{-1}$ is the spatial gradient of the "inverse motion" (*cf.* Chapter 2), and N is the entropy per unit reference volume. As a matter of fact, we more precisely have

$$H = \frac{1}{2\,\rho_0} \mathbf{P} \cdot \mathbf{C}^{-1} \cdot \mathbf{P} + E(\mathbf{X}, \mathbf{F}^{-1}, N), \tag{4.63}$$

where ρ_0 is the reference (possibly $\mathbf{X}$ dependent) matter density, E is the *internal energy* per reference volume, $\mathbf{P} = \rho_0\,\mathbf{C} \cdot \mathbf{V}$, $\mathbf{C} = \mathbf{F}^T \cdot \mathbf{F}$, and $\mathbf{V} = -\mathbf{F}^{-1} \cdot \mathbf{v}$, with $\mathbf{v} = \partial\bar{\mathbf{x}}/\partial t$, $\mathbf{F} = \nabla_R\bar{\mathbf{x}}$. This is not the standard formulation of elasticity, but the one using the so-called "inverse motion" (see Chapter 2). We let the reader do the tricky exercise that the second of (4.61) then yields the canonical equations of energy and momentum in the form:

$$\frac{dN}{dt} = 0, \qquad \frac{d\mathbf{P}}{dt} = div_R\,\mathbf{b} + \mathbf{f}^{\text{inh}}, \tag{4.64}$$

wherein

$$\mathbf{b} = -(L\,\mathbf{1} + \mathbf{T} \cdot \mathbf{F}), \quad L = \mathbf{P} \cdot \mathbf{V} - H = K - E,$$

$$\mathbf{T} = \frac{\partial\overline{E}}{\partial\mathbf{F}}, \qquad\qquad \mathbf{f}^{\text{inh}} = \left.\frac{\partial L}{\partial\mathbf{X}}\right|_{\text{expl}}. \tag{4.65}$$

The conservation-like equation for the *action* is then obtained by applying Noether's theorem for a group of *simultaneous* space and time transformations (expansions or *scaling*) as shown by one of the authors (Maugin (2011b, Chapter 4)) and also Lazar and Anastassiadis (2007).

Remark 4.1. Conservation of wave momentum. In the 1D linear case, because of $D_L(\omega, K) = 0$, $\widetilde{L}$ is originally quadratic in ω, and thus Equation (4.40)$_1$ yields $\widetilde{P} = \omega\,K\,\mathbf{a}^2$. This is also the case of the standard wave momentum $p^w = -u_t\,u_x = \omega\,K\,\mathbf{a}^2$ *modulo* a factor of modulus one. They are thus identical up to this factor.

4.7 The case of inhomogeneous waves

In a thoughtful series of papers W. D. Hayes (1970a,b, 1973, 1974b) proposed a viewpoint slightly different from that of Whitham. He considered fields ϕ^α, $\alpha = 1, 2, \ldots$ that depend on $\mathbf{x}$, t and a phase φ. They are of period 2π in φ. The Lagrangian density reads

$$L = L(\dot{\phi}^\alpha, \nabla\phi^\alpha, \phi^\alpha, \mathbf{x}, t). \tag{4.66}$$

The corresponding field equations (Euler–Lagrange equations for each α) are obtained as

$$\frac{\partial}{\partial t}\left(\frac{\partial L}{\partial \dot{\phi}^\alpha}\right) + \nabla \cdot \left(\frac{\partial L}{\partial \nabla\phi^\alpha}\right) - \frac{\partial L}{\partial \phi^\alpha} = 0. \tag{4.67}$$

We introduce again the following notation for the average of a field over the period 2π:

$$\tilde{f}(\mathbf{x}, t) = \langle f(\mathbf{x}, t, \varphi)\rangle := \frac{1}{2\pi}\int_0^{2\pi} f(\mathbf{x}, t, \varphi)\, d\varphi. \tag{4.68}$$

The action density and action flux are then defined by

$$A(\mathbf{x}, t) := \left\langle \frac{\partial L}{\partial \dot{\phi}^\alpha}\frac{\partial \phi^\alpha}{\partial \varphi}\right\rangle, \tag{4.69}$$

and

$$\mathbf{B}(\mathbf{x}, t) := \left\langle \frac{\partial L}{\partial \nabla\phi^\alpha}\frac{\partial \phi^\alpha}{\partial \varphi}\right\rangle, \tag{4.70}$$

where summation over α is understood. The *local conservation law of action*

$$\frac{\partial A}{\partial t} + \nabla \cdot \mathbf{B} = 0, \tag{4.71}$$

is proved by multiplying each of (4.67) by the corresponding $\partial\phi^\alpha/\partial\varphi$, summing over α, taking the average of the result over the period of φ, and noting that

$$\left\langle \frac{\partial L}{\partial \varphi}\right\rangle = \frac{1}{2\pi}\int_0^{2\pi}\frac{\partial L}{\partial \varphi}\, d\varphi \equiv 0. \tag{4.72}$$

Hayes first envisages a wave motion for which L depends explicitly neither on t nor on the propagation space $\mathbf{x}_\parallel$ and strictly periodic solutions are expected of period 2π in the new phase variable

$$\overline{\varphi} = \mathbf{k}\cdot\mathbf{x}_\parallel - \omega t - \varphi. \tag{4.73}$$

This provides waves that are strictly *plane* in propagation space $\mathbf{x}_\|$. The notation $\mathbf{x}_{\mathrm{cr}}$ will indicate the *cross* or *lateral* space. But more generally, just as with Whitham (see Appendix B), introducing slow space and time variable by $\mathbf{X} = \varepsilon\,\mathbf{x}_\|$ and $T = \varepsilon\,t$ and a general phase function

$$\overline{\varphi} = \overline{\varphi}(\mathbf{x}_\|, t), \tag{4.74}$$

we define a frequency ω and a wave vector $\mathbf{k}$ by (*cf.* Eq. (4.29))

$$\omega(\mathbf{X}, T) = -\frac{\partial \overline{\varphi}}{\partial t}, \qquad \mathbf{k}(\mathbf{X}, T) = -\nabla_\| \overline{\varphi}, \tag{4.75}$$

so that we have the consistency relation (conservation of wave vector in propagation space)

$$\frac{\partial \mathbf{k}}{\partial t} + \nabla_\| \omega = 0. \tag{4.76}$$

Hayes call wave solutions of the *local type* those observed when the cross or lateral space is reduced to a point. When the cross space is not so trivial, the corresponding waves are called *local* waves. Here we must distinguish between the case when homogeneous boundary conditions apply at a fixed boundary in cross space or appropriate evanescence conditions apply if this cross space is not bounded (this will be the case of surface waves in a half space) in which case the wave solutions are called *modal*, or when such conditions are either partially applied or not at all, in which case the wave solutions are called *general* ones. The first of these two cases is of interest because it yields an eigenvalue-eigenfunction problem of which the solution does not exist for arbitrarily specified ω, $\mathbf{k}$ and amplitude.

For solutions $\phi^\alpha(\mathbf{x}_{\mathrm{cr}}, \overline{\varphi})$ in the *modal* case, with appropriate homogeneous boundary conditions or evanescence conditions at infinity, the following result holds true:

$$\int_{\mathbf{x}_{\mathrm{cr}}} \left\langle \frac{\partial L}{\partial \dot\phi^\alpha}\,\dot\phi^\alpha + \frac{\partial L}{\partial \nabla \phi^\alpha}\,\nabla \phi^\alpha + \frac{\partial L}{\partial \phi^\alpha}\,\phi^\alpha \right\rangle d\mathbf{x}_{\mathrm{cr}} = 0. \tag{4.77}$$

Also, as both A and $\mathbf{B}$ do not depend explicitly on t and $\mathbf{x}_\|$, it is also shown that

$$\nabla_{\mathrm{cr}} \cdot \mathbf{B}_{\mathrm{cr}} = 0. \tag{4.78}$$

[Whenever L is quadratic in its three arguments, because of Euler identity, Equation (4.77) in fact means that

$$\int_{\mathbf{x}_{\mathrm{cr}}} \langle L \rangle\, d\mathbf{x}_{\mathrm{cr}} = 0. \tag{4.79}$$

In particular, this holds true in linear elasticity where the ϕ^α's are none other than the components of the elastic displacement and L does not depend on this displacement because of Galilean invariance.]

We can define an action density $\widetilde{A}$ and action flux $\widetilde{\mathbf{B}}$ in propagation space by

$$\widetilde{A} = \int_{\mathbf{x}_{\mathrm{cr}}} A \, d\mathbf{x}_{\mathrm{cr}}, \qquad \widetilde{\mathbf{B}} = \int_{\mathbf{x}_{\mathrm{cr}}} \mathbf{B} \, d\mathbf{x}_{\mathrm{cr}}. \tag{4.80}$$

Then for solutions close to strictly periodic ones in a uniform medium, we will have a conservation law of action in propagation space as

$$\frac{\partial \widetilde{A}}{\partial t} + \nabla_\parallel \cdot \widetilde{\mathbf{B}} = 0, \tag{4.81}$$

with [compare to (4.37)]

$$\widetilde{A} = \frac{\partial \widetilde{L}}{\partial \omega} = \widetilde{S}, \qquad \widetilde{\mathbf{B}} = -\frac{\partial \widetilde{L}}{\partial \mathbf{k}} = \widetilde{\mathbf{W}}. \tag{4.82}$$

Note. Quantization of elastic deformation field. Although in this book we are not really concerned with quantum mechanics, the curious reader may be interested by the following remark about the quantization of crystal vibrations. We consider the total wave momentum

$$P^w = -\int_V \rho \, u_t \, u_x \, dV. \tag{4.83}$$

As we know, quantum mechanics replaces functions by *operators*. With conjugated Hermitian operators $a_{\mathbf{k}}$ and $a_{\mathbf{k}}^+$ — associated with the $\mathbf{k}$ wave vector — which satisfy the commutation rules

$$\left[a_{\mathbf{k}}, a_{\mathbf{k}'}^+\right] = \delta_{\mathbf{k}\mathbf{k}'}, \qquad \left[a_{\mathbf{k}}, a_{\mathbf{k}'}\right] = \left[a_{\mathbf{k}}^+, a_{\mathbf{k}'}^+\right] = 0, \tag{4.84}$$

with $[A, B] = AB - BA$, and introducing phonon creation and annihilation operators $b(\mathbf{k})$ and $b^+(\mathbf{k})$ by

$$a_{\mathbf{k}} = (2\pi/L)^{3/2} \, b(\mathbf{k}), \qquad a_{\mathbf{k}}^+ = (2\pi/L)^{3/2} \, b^+(\mathbf{k}), \tag{4.85}$$

while the Kronecker symbol $\delta_{\mathbf{k}\mathbf{k}'}$ is replaced by a delta function $\delta(\mathbf{k} - \mathbf{k}')$ such that

$$\delta_{\mathbf{k}\mathbf{k}'} \to (2\pi)^3 \, \delta(\mathbf{k} - \mathbf{k}')/V, \tag{4.86}$$

with $V = L^3$, it is shown (Kosevich, 1999, Section 6.6) that (4.1) yields the operator expression

$$P^w = \hbar \sum_{\mathbf{k}} \mathbf{k} \, a_{\mathbf{k}}^+ a_{\mathbf{k}} = \hbar \int \mathbf{k} \, b^+(\mathbf{k}) \, b(\mathbf{k}) \, d^3\mathbf{k}, \tag{4.87}$$

so that the total wave momentum P^w is the total *quasi-momentum* of a vibrating crystal. For phonon interactions and anharmonicity of crystal vibrations (nonlinear effects), one must consider a phonon gas that is no longer ideal (*cf.* Kosevich, 1999, Chapter 7). This is outside the scope of the present book, although of great interest. More on quasi-momentum in elasticity, light and matter is to be found in Gurevich and Thellung (1990, 1992).

Chapter 5

Transmission-reflection problem

Object of the chapter

Here the emphasis is placed on the possible exploitation of the wave-quasi-particle dualism in the classical problem of the transmission and reflection of waves by a discontinuity between two media (in perfect contact or with possible delaminating) and the more general case of a mono-layer or multi-layer sandwiched slab. More precisely, this standard propagation problem (of interest in nano-systems and seismic studies) considers only one elastic component of the SH (shear-horizontal) type and the normal incidence of such a wave on interfaces between elastic continua characterised by their own elastic (shear) coefficient and mass density. The resulting quasi-particles exhibit a Newtonian dynamics.

5.1 Introduction

In this chapter, relying on a simple case of wave — one elastic component of the SH (shear-horizontal) type —, and considering the normal incidence of such a wave on the interface between two elastic continua characterised by their own elastic (shear) coefficient and mass density, we revisit the problem of the transmission and reflection of such a wave at this interface in terms of the wave properties (this is only a reminder of well known results) or those of the associated quasi-particle properties (this is the new contribution). The situation considered is that which prevails in the potential exploitation in nanoscale systems. For macro-systems this would be the realm of potentially dangerous seismic waves. In the present case, discarding the variation of amplitude with depth we are satisfied with considering a plane face wave. The representative volume element of the wave modes then is one wavelength in the propagation direction and a unit square in the transverse plane.

5.2   Reminder on the wavelike picture

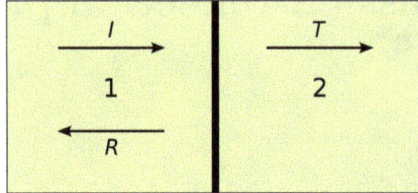

Fig. 5.1   The transmission-reflection problem at an interface.

5.2.1   *One-dimensional case*

For the sake of simplicity we consider a one-dimensional wave problem along the x-direction (Figure 5.1) for an elastic displacement u that may be a transverse shear-horizontal (SH) mode. The governing wave equation from Chapter 2 is given by

$$\frac{\partial p}{\partial t} - \frac{\partial \sigma}{\partial x} = 0, \qquad p = \rho \frac{\partial u}{\partial t}, \qquad \sigma = \mu \frac{\partial u}{\partial x}, \qquad (5.1)$$

where ρ and μ are a prescribed matter density and fixed elasticity (e.g., shear) coefficient μ in a given region of space. Space and time partial derivatives are alternately denoted by a comma followed by x or t. The first of (5.1) can be deduced in a variational format from the following Lagrangian density per unit volume

$$L = \frac{1}{2} \rho u_{,t}^2 - \frac{1}{2} \mu u_{,x}^2, \qquad (5.2)$$

We consider a harmonic wave motion

$$u = U \cos(k x - \omega t), \qquad (5.3)$$

so that from (5.1) we have the trivial dispersion relation

$$D(\omega, k) = \omega^2 - c^2 k^2 = 0, \qquad c = (\mu/\rho)^{1/2}. \qquad (5.4)$$

5.2.2 Transmission-reflection problem for a perfect interface

The standard transmission-reflection problem here consists in considering a normally incident wave in medium 1 (properties ρ_1 and μ_1) on the interface at $x = x_0 = 0$ that separates medium 1 from medium 2 (properties ρ_2 and μ_2). The wave solution in medium 1 is sought as the sum of an incident component and a reflected component while the solution in medium 2 consists in a transmitted component. With an obvious notation we have thus

$$u_1 = u_I + u_R, \qquad u_2 = u_T, \tag{5.5}$$

with

$$
\begin{aligned}
u_1 &= U \cos(k_1 x - \omega t) + R_0 U \cos(k_1 x + \omega t), \\
u_2 &= T_0 U \cos(k_2 x - \omega t),
\end{aligned}
\tag{5.6}
$$

where R_0 and T_0 are the reflection and transmission coefficients for this *perfect interface* case for which we assume the continuity of displacement and stress, i.e.,

$$u_1 = u_2, \quad \mu_1 u_{1,x} = \mu_2 u_{2,x} \qquad \text{at } x = 0. \tag{5.7}$$

The solution of this system is, for any amplitude U, the value of the reflection and transmission coefficients as (Brekhovskikh (1960))

$$R_0 = \frac{z_1 - z_2}{z_1 + z_2}, \qquad T_0 = \frac{2 z_1}{z_1 + z_2}, \tag{5.8}$$

where $z_\alpha = \rho_\alpha c_\alpha$, $\alpha = 1, 2$ are impedances. With these we check the conservation of energy flux in the well known form

$$F_0 = 1 - R_0^2 - \frac{z_2}{z_1} T_0^2 \equiv 0. \tag{5.9}$$

5.2.3 Transmission-reflection problem for an interface with delamination

In view of applications to non-destructive (NDT) evaluation methods, it is of interest to consider the case of an imperfect interface that admits the possibility of delaminating and for which the matching conditions (5.7) are replaced by the conditions (known as Jones' conditions (Jones and Whittier (1967))

$$\sigma_1 = \sigma_2 \equiv K [u], \tag{5.10}$$

where K is a positive coefficient characterizing the degree of delaminating and the symbol $[..]$ means the jump of its enclosure, i.e., $[u] = u_2 - u_1$ at $x = 0$. We must look for *complex* solutions of the type $u = A \exp\left(i\left(k\,x - \omega\,t\right)\right)$. The conditions (5.10) yield the system

$$z_1\left(1 - R\right) = z_2\,T, \qquad z_2\,T = -i\,\frac{K}{\omega}\left(1 + R - T\right), \qquad (5.11)$$

where R and T are the *complex* reflection and transmission coefficients corresponding to this imperfect case.

The solution of (5.11) reads

$$R = \frac{z_1\,z_2 - i\left(K/\omega\right)\left(z_1 - z_2\right)}{z_1\,z_2 - i\left(K/\omega\right)\left(z_1 + z_2\right)}, \qquad (5.12)$$

and

$$T = \frac{-2\,i\left(K/\omega\right)z_1}{z_1\,z_2 - i\left(K/\omega\right)\left(z_1 + z_2\right)}. \qquad (5.13)$$

By computing the square of the moduli of the complex quantities R and T, we check that (5.9) is replaced by

$$F_K = 1 - |R|^2 - \frac{z_2}{z_1}\,|T|^2 \equiv 0. \qquad (5.14)$$

We note that

$$F_K = F_0\left(1 - \frac{z_1^2\,z_2^2}{z_1^2\,z_2^2 + (K/\omega)^2\,(z_1 + z_2)^2}\right). \qquad (5.15)$$

The solution of this imperfect interface case is characterized by the parameter K/ω which shows the role played by the frequency ω. The limit case $K \to \infty$ corresponds to the perfect interface for which (5.9) holds true. The limit case $K \to 0$ corresponds to full delaminating (no more transmission and complete reflection: $T = 0$, $R = 1$).

It is possible to obtain an estimate of the K coefficient by measuring the amplitude of reflected or transmitted waves. In particular, whenever media 1 and 2 are identical, but K still is not zero, this measure provides a means of determining the presence of an internal delaminating in the body.

5.3   Associated quasi-particle picture

5.3.1   *Basic equations*

With the field equation (5.1) there are associated (*via* the application of Noether's theorem or by direct manipulation) conservation laws of energy and wave momentum in the local form in a homogeneous medium (*cf.*, Section 3.5):

$$\frac{\partial H}{\partial t} - \frac{\partial Q}{\partial x} = 0, \qquad \frac{\partial P}{\partial t} - \frac{\partial b}{\partial x} = 0, \tag{5.16}$$

where the energy or Hamiltonian per unit volume H, the energy flux Q, the wave momentum P and the (here reduced to a scalar) Eshelby stress b are defined by (see Chapter 3 and Appendix A for the canonical definitions in three dimensions)

$$H = E + W = \frac{1}{2}\rho\,u_{,t}^2 + \frac{1}{2}\mu\,u_{,x}^2, \quad Q = \sigma\,u_t = \mu\,u_{,x}\,u_{,t}, \tag{5.17}$$

$$P = -\rho\,u_{,t}\,u_{,x}, \qquad b = -(L + \sigma\,u_x). \tag{5.18}$$

The first of these last two follows Brenig's (1955) definition. It is a remarkable — but sometimes misleading — fact that in this one-dimensional case the following identities hold

$$Q = -c^2\,P, \qquad b = -H. \tag{5.19}$$

Now the concept of quasi-particle is introduced by integrating the conservation equations (5.16) over a volume that is representative of the present wave motion, i.e., over one wavelength in the x propagation direction and a square section of sides equal to unity in the transverse direction. This introduces averages noted with the symbolism $\langle .. \rangle$. For solutions of the type (5.3) this procedure yields the following results:

$$\langle L \rangle = 0, \tag{5.20}$$

$$\frac{d}{dt}\langle P \rangle = 0, \tag{5.21}$$

$$\frac{d}{dt}\langle H \rangle = 0, \tag{5.22}$$

where, in a homogeneous medium,

$$\langle P \rangle = \rho\,\omega\,\pi\,U^2 \equiv M\,c, \qquad \langle H \rangle = \frac{1}{2}\,M\,c^2, \tag{5.23}$$

where the first of these defined the "mass" $M = \rho\,k\,\pi\,U^2$. Simultaneously, equations (5.21) and (5.22) show that the said quasi-particle has an inertial Newtonian motion. It also satisfies the Lebnizian conservation of kinetic energy (or *vis-viva*). Indeed, while there are both kinetic and elastic energy at the continuum level, the energy of the associated quasi-particle is purely *kinetic* on account of the given definition of "mass". Note that the homogeneous equations (5.21) and (5.22) hold good because of the periodicity at the two ends of the integration interval that makes the contributions from Q and b to vanish. Equation (5.20) holds true because it is shown in computing the average for solutions (5.3) that the result is proportional to the "dispersion" relation (5.4) and therefore vanishes identically, a result known in the kinematic-wave mechanics of Lighthill and Whitham (see, e.g., Whitham (1974)). Note finally that all quantities introduced in (5.20)–(5.23) are proportional to the square of the amplitude of the wave, hence are all of energetic nature.

5.3.2 *Transmission-reflection problem (perfect interface)*

Considering first the case of the perfect interface at $x = 0$, we can associate one quasi-particle with each wave component of the problem. With an obvious notation we have the following "masses":

$$M_I = \rho_1 \, k_1 \, \pi \, U^2, \quad M_R = \rho_1 \, k_1 \, \pi \, R_0^2 \, U^2, \quad M_T = \rho_2 \, k_2 \, \pi \, T_0^2 \, U^2. \quad (5.24)$$

The corresponding averaged wave momenta are given by

$$\overline{P}_I \equiv \langle P_I \rangle = \rho_1 \, \omega \, \pi \, U^2, \quad (5.25)$$

and

$$\overline{P}_R \equiv \langle P_R \rangle = -\rho_1 \, \omega \, \pi \, R_0^2 \, U^2, \qquad \overline{P}_T \equiv \langle P_T \rangle = \rho_2 \, \omega \, \pi \, T_0^2 \, U^2, \quad (5.26)$$

where we account for the fact that the averaged wave momentum $\overline{P}_R$ is oriented towards negative x's.

We note ΔM and $\Delta \overline{P}_R$ the possible misfits in mass and momentum defined by

$$\Delta M := (M_R + M_T) - M_I, \quad (5.27)$$

and

$$\Delta \overline{P} := (|\overline{P}_R| + |\overline{P}_T|) - |\overline{P}_I|, \quad (5.28)$$

where the symbolism $|\dots|$ refers to the absolute value of its enclosure. That is, we are comparing the strengths of the momenta and, therefore, we are not performing a vectorial balance.

Similarly for the kinetic energy of the associated quasi-particles:

$$\Delta \overline{H} = (\overline{H}_R + \overline{H}_T) - \overline{H}_I. \quad (5.29)$$

We say that a quantity is conserved during the transmission-reflection problem if the corresponding misfit vanishes.

By applying the definition $(5.23)_2$, one immediately shows that

$$\Delta\overline{H} = \left(\frac{1}{2} z_1 \,\omega\, \pi\, U^2\right) F_0, \qquad (5.30)$$

where F_0 has been defined in (5.9). But the latter vanishes. Accordingly, $\Delta\overline{H} \equiv 0$: *kinetic energy is conserved* in the transmission-reflection problem seen as a quasi-particle process that may be qualified of Leibnizian (conservation of *vis-viva*). Historically, it was in fact the collision problem between particles that led the late 17th and early 18th century scientists (prominently Leibniz in 1686; see Smith (2006) to introduce the *vis-viva* — twice the actual kinetic energy — as the "physical quantity" to be conserved in such interactions and not Descartes' "quantity of motion" as suggested by a different school (see below for the particle momentum in the present problem).

Indeed, "mass" is not generally conserved in the present problem as it is immediately shown that (Maugin and Rousseau, 2012b)

$$\Delta M = \rho_2 \, k_2 \, \pi \, T_0^2 \, U^2 \left(\frac{c_1^2 - c_2^2}{c_1^2}\right). \qquad (5.31)$$

Similarly, on computing $\Delta\overline{P}$ it is obtained that

$$\Delta\overline{P} = \rho_2 \, k_2 \, \pi \, T_0^2 \, U^2 \left(\frac{c_1 \, c_2 - c_2^2}{c_1}\right). \qquad (5.32)$$

This shows that

$$\Delta\overline{P} = \left(\frac{c_1 \, c_2}{c_1 + c_2}\right) \Delta M. \qquad (5.33)$$

So $\Delta\overline{P}$ and ΔM always are in the same sign. In particular, $\Delta M > 0$ if $c_1 > c_2$ and $\Delta M < 0$ if $c_1 < c_2$ ($\Delta M = 0$ if and only of $c_1 = c_2$ that is, if there is no interface or for the extraordinary case where the two media have elasticity coefficients and densities in the same ratio). The momentum $\Delta\overline{P}$ vanishes in the same conditions. So that, $\overline{P}$ and M are conserved when media 1 and 2 have the same characteristic elastic speed, when medium 2 is a vacuum ($\rho_2 \to 0$) and when the medium 2 is perfectly rigid ($\mu_2 \to \infty$). In these cases, there is perfect reflection ($R_0^2 = 1$) from the wavelike viewpoint or perfect (i.e., elastic) rebound from the particle-like viewpoint.

The result obtained for a normal incidence can be generalized to an oblique incidence. Computations then show that now there obviously exists a vectorial misfit in the wave momenta, but this results solely from the misfit in the normal component.

5.3.3   Case of an imperfect interface for an interface with delaminating

Remarkably enough there is no need to do again the computations. It suffices to replace the transmission and reflection coefficients of the perfect case by the moduli of the new complex coefficients.

Thus, (5.30) is replaced by

$$\Delta \overline{H} = \left(\frac{1}{2} z_1 \, \omega \, \pi \, U^2 \right) F_K , \qquad (5.34)$$

and this again vanishes. Similarly, (5.31) and (5.32) hold with T_0^2 replaced by $|T|^2$ while (5.33) remains unchanged, noting that the coefficient $c_1 c_2 / (c_1 + c_2)$ does not depend on K.

In the case when $K \neq 0$ but media 2 and 1 are identical, the presence of the K spring distribution can simulate a homogeneous surface damage. In this case, both ΔM and $\Delta \overline{P}$ vanish so that K is no longer involved. The dependence on K shows only through the value of any of M_R, M_T, $\overline{P}_R$ and $\overline{P}_T$.

In the case when $K \neq 0$ but media 1 and 2 are different, both ΔM and $\Delta \overline{P}$ do not vanish. This result is not due to the presence of the K spring distribution; it can only be attributed to the difference in characteristic wavelength in both media. These wavelengths depend on the integration length of propagation in each medium. This hypothesis being accepted, the value of K can also be evaluated from ΔM and $\Delta \overline{P}$. As the surface damage increases (i.e., as K decreases), the mass of the reflected quasi-particle increases and that of the transmitted quasi-particle decreases compared to the fully undamaged case.

5.4   Case of a sandwiched slab

Here we consider the case of an elastic slab (medium 2) of thickness d situated between $x = 0$ and $x = d$ and sandwiched between two media of elastic type 1, propagation is from left to right with reflection coefficient R in the left medium 1 and transmission coefficient T in the right medium 1. We assume that $d \gg \lambda_2$, where λ_2 is the (elastic wave) characteristic wavelength of medium 2, so that the association of quasi-particle properties makes sense in the slab. We need not reconsider the wavelike solution. We are satisfied with applying the results of the foregoing section to the two interfaces at $x = 0$ (transition $1 \rightarrow 2$) and at $x = d$ (transition $2 \rightarrow 1$).

Just as before $(M_I, M_R, M_T, P_I, P_R, P_T)$ are the masses and the momenta granted to the quasi-particles in left and right regions 1. We obviously have (compare (5.24))

$$M_I \propto U^2, \quad M_R \propto |R|\, U^2, \quad M_T \propto |T|\, U^2, \quad |R|^2 + |T|^2 = 1, \quad (5.35)$$

and in an obvious notation

$$\Delta M_{11} = (M_R + M_T) - M_I = 0. \tag{5.36}$$

As to the momenta, we have (*cf.* Section 5.3)

$$\Delta \overline{P}_{11} = (|P_R| + |P_T|) - |P_I| \equiv 0. \tag{5.37}$$

Within the slab we distinguish between the particle momentum P^+ with mass M^+ associated with the right motion and particle momentum P^- with mass M^- associated with the left motion.

Thus for the interface $x = 0$ we can write

$$\begin{aligned} \Delta M_{1\to 2} &:= (M^+ + M^- + M_R) - M_I \\ &= (M^+ + M^-) - M_T \neq 0. \end{aligned} \tag{5.38}$$

and

$$\begin{aligned} \Delta \overline{P}_{1\to 2} &= (|P^+| + |P^-| + |P_R|) - |P_I| \\ &= (|P^+| + |P^-|) - |P_T| \neq 0, \end{aligned} \tag{5.39}$$

while for the interface $x = d$, we have similarly

$$\Delta M_{2\to 1} = M_T - (M^+ + M^-) = -\Delta M_{1\to 2} \neq 0, \tag{5.40}$$

and

$$\Delta \overline{P}_{2\to 1} = |P_T| - (|P^+| + |P^-|) = -\Delta \overline{P}_{1\to 2} \neq 0. \tag{5.41}$$

Note that

$$(|P^+| + |P^-|) = \Delta \overline{P}_{1\to 2} + |P_T|, \tag{5.42}$$

and obviously,

$$R \equiv R_{1\to 2\to 1}, \quad T \equiv T_{1\to 2\to 1}. \tag{5.43}$$

These two coefficients can be computed in terms of the mechanical properties of media 1 and 2. A standard calculation involving the matching conditions at the crossing of the two interfaces separated by the distance d yields

$$T = \frac{1}{\cos(k_2\, d) - \frac{i}{2}\left(\frac{z_1}{z_2} + \frac{z_2}{z_1}\right)\sin(k_2\, d)}, \tag{5.44}$$

and

$$R = \frac{-\frac{i}{2}\left(\frac{z_1}{z_2} - \frac{z_2}{z_1}\right)\sin(k_2\,d)}{\cos(k_2\,d) - \frac{i}{2}\left(\frac{z_1}{z_2} + \frac{z_2}{z_1}\right)\sin(k_2\,d)}. \tag{5.45}$$

It is readily checked that these two formulas comply with the conservation law $|R|^2 + |T|^2 = 1$. Of course we also check that $T = 1$ and $R = 0$ if $d \equiv 0$ (perfect contact). As to the condition $d \ll \lambda_2$, it does not make sense for the introduction of a quasi-particle in the slab.

The type of formalism, book keeping and "algebra" just introduced can be applied to the more complicated case where the sandwiched slab is made of a number $n - 1$ of perfectly elastic layers (each with its own elastic properties) numbered $i = 2, \ldots, n$, in perfect contact. That is, the medium left to the slab and the medium right to it are made of the same material (material 1). The estimates in (5.35) still hold good. This is also the case of (5.36). Same as in (5.38), we will write

$$\Delta M_{1\to 2} = (M_2^+ + M_2^- + M_R) - M_I \neq 0, \tag{5.46}$$

$$\Delta M_{2\to 3} = (M_3^+ + M_3^-) - (M_2^+ + M_2^-) \neq 0, \tag{5.47}$$

and so on,

$$\Delta M_{i\to i+1} = (M_{i+1}^+ + M_{i+1}^-) - (M_i^+ + M_i^-) \neq 0, \tag{5.48}$$

$$\Delta M_{n-1\to n} = (M_n^+ + M_n^-) - (M_{n-1}^+ + M_{n-1}^-) \neq 0, \tag{5.49}$$

$$\Delta M_{n\to 1} = M_T - (M_n^+ + M_n^-) \neq 0, \tag{5.50}$$

Here superscripts $+$ and $-$ indicate masses associated with particle motions to the right and the left, respectively. On summing over all equations (5.46) through (5.50) we re-obtain (5.36). A similar book keeping can be done for the momentum transfers $\Delta \overline{P}_{11}$ (*cf.* equation (5.37)) and $\Delta \overline{P}_{i\to i+1}$. We also note that, globally,

$$R = R_{1\to\ldots\ldots\to 1}, \qquad T = T_{1\to\ldots\ldots\to 1}, \tag{5.51}$$

where the dots stand for the $(n-1)$ layers numbered $(2, \ldots, n)$. In principle the global reflection and transmission coefficients (5.51) can be computed in terms of the individual properties of the layers, but this is not of an immediate interest.

5.5   Conclusion

The above reported analysis shows that, in the transmission-reflection problem within the framework of purely elastic material regions, the resulting mechanics of the associated quasi-particles results in an exchange of linear momentum without any global loss of it while there also is no misfit in "mass" but the masses depend on the region of propagation, i.e., the various propagation properties.

Interestingly enough, there is no misfit in the kinetic energies of the quasi-particles — *cf.* equations (5.30) and (5.34). This is a direct consequence of the acoustic (wavelike) energy conservation represented by equations (5.9) and (5.14). There is no principle difficulty in dealing with the slab problem and its extension to a multi-layered medium. But there may be computational difficulties. The case of delaminating at an interface fits in the picture and the properties of this delaminating (spring coefficient K) can be deduced from the "misfits" in mass and momentum of the associated quasi-particles.

One may wonder what happens if some of the involved media of propagation are viscoelastic. The wavelike problem of an interface between two media was studied a long time ago (see Mackenzie (1960), and references therein), at least for a small perturbing viscosity. Insofar as the concept of associated quasi-particle is concerned, for a propagation of surface SH waves in an infinite continuum along the propagation direction we have shown elsewhere (Rousseau and Maugin (2012a)) the complexity and originality of the result. Not only does the viscous damping result in a friction force as a source in the equation of motion of the associated quasi-particle, but the "mass" itself of this so-called particle is found to vary in time. One can easily imagine the analytical difficulties to be met in treating the case of a viscoelastic slab sandwiched between two identical elastic or viscoelastic media. The limit case of the slab thickness going to zero should provide the case of an absorbing interface of vanishing thickness thus re-instating the notion of coefficient of restitution.

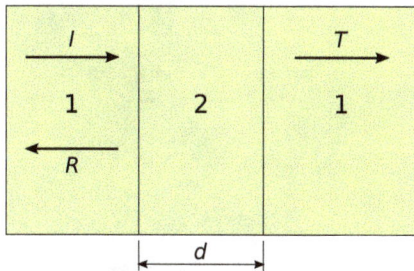

Fig. 5.2   Transmission-reflection problem through an embedded slab.

Chapter 6

Application to dynamic materials

Object of the chapter

So-called dynamic materials with their smoothly or abruptly varying properties in both space and time are materials that lend themselves very well to the study of consequences of inhomogeneities, the related conservation laws of energy and wave momentum, and the associated quasi-particles. This matter here is duly examined with a comparison between the traditional wave-like picture and the more original re-interpretation in terms of the motion of quasi-particles. The latter proves to be more easily physically grasped in several instances.

6.1 Reminder on the notion of dynamic materials

We are interested in so-called *dynamic materials*. By these we understand materials of which the characteristic properties (in the simplest case of elastic bodies, mass density and elasticity) may be made to vary in space or in time, or both, by an appropriate arrangement or control.

Examples of appropriate governing equations in one spatial dimension are given in Section 3.5. We obviously ignore here inhomogeneous materials with stochastic properties. Materials inhomogeneous in space are plenty, mostly illustrated by composite materials or structures made of tightly assembled bodies of differing mechanical properties. Materials inhomogeneous in time are not so frequent or are practically nonexistent in everyday conditions (room temperature, time scale in minutes or hours). We may conceive of some artificial means of causing these controlled changes in time, for instance, by the application of an external (non-mechanical) field, or through a phase transition. To avoid any misunderstanding, we specify that this should be realized in an infinitesimally short duration and over a sufficiently large material region if not over the whole specimen under consideration.

The idea of dynamic materials was formulated by Blekhman and Lurie (2000) who identified two kinds of dynamic materials: *activated dynamic materials* obtained by changing the material coefficients of the wave carrier medium in the absence of relative motion — the subject matter of the present chapter — and *kinetic dynamic materials* obtained by endowing the whole system or some of its regions with some prescribed relative motion. The book of Lurie (2007) concerns the first type with applications mostly to electromagnetic materials. The second type in mechanics is best illustrated by the review of Vesnitskii and Metrikine (1996) — and the many works of the Nizhny–Novgorod school referred to in that lengthy and detailed paper. That latter work emphasizes the interest to study so-called *transition radiation* in mechanical systems such as elastic waves generated by mechanical objects travelling in locally or periodically inhomogeneous elastic systems (e.g., train travelling on a railroad track supported by a more or less elastic ground) while recalling the relationship and differences with *transition radiation* in some electromagnetic systems previously studied in detail by Ginzburg and Tsytovich (1979), although Vesnitski and Metrikine do not consider wave phenomena such as Cherenkov or bremsstrahlung radiation. The whole problem, however, presents analytical difficulties that cannot be overlooked. From the mathematical side, we note a few works on the existence of special solutions of wave systems that are inhomogeneous in both space and time (e.g., Nadin (2009) and other works by the same author). This obviously relates to the most mathematical facets of our problem.

In what follows, in order to avoid combining complexities from various sides, the following reported considerations concern only one-dimensional systems in space as illustrated in Section 3.5. For such a system, with a standard notation (u = elastic displacement; derivatives indicated by a subscript x or t), the Lagrangian density reads

$$L = \frac{1}{2}\,\hat{\rho}_0(x,t)\,(u_t)^2 - \frac{1}{2}\,\widehat{E}(x,t)\,(u_x)^2. \tag{6.1}$$

The equation of linear momentum takes the form

$$\frac{\partial}{\partial t}\left(\hat{\rho}_0(x,t)\,\frac{\partial u}{\partial t}\right) - \frac{\partial}{\partial x}\left(\widehat{E}(x,t)\,\frac{\partial u}{\partial x}\right) = 0. \tag{6.2}$$

The canonical quantities of Hamiltonian density H, energy flux Q, wave momentum p^w, and Eshelby stress b now are all scalars such that (*cf.* Chapter 3)

$$H = \frac{1}{2}\,\hat{\rho}_0(x,t)\,(u_t)^2 + \frac{1}{2}\,\widehat{E}(x,t)\,(u_x)^2, \tag{6.3}$$

$$Q = -u_t \frac{\partial L}{\partial u_x} = \widehat{E}\, u_t\, u_x, \tag{6.4}$$

$$p^w = -u_x \frac{\partial L}{\partial u_t} = -\hat{\rho}_0\, u_t\, u_x, \tag{6.5}$$

$$b = -\left(L - u_x \frac{\partial L}{\partial u_x}\right) = -H. \tag{6.6}$$

We recall that the latter result is a peculiarity due to the one-dimensionality of the problem in space.

The canonical local equations of energy and wave momentum read [*cf.* Equations (3.50) and (3.51)]

$$\frac{\partial H}{\partial t} - \frac{\partial Q}{\partial x} = h := \left.\frac{\partial L}{\partial t}\right|_{\text{expl}} \tag{6.7}$$

and

$$\frac{\partial p^w}{\partial t} - \frac{\partial b}{\partial x} = f := \left.\frac{\partial L}{\partial x}\right|_{\text{expl}}. \tag{6.8}$$

The special forms of h and f are immediately read off from (6.1).

To simplify further the problem we shall consider the following special case of inhomogeneities in space and time:

$$\rho_0 = \hat{\rho}_0(x), \qquad W = \widehat{W}(u_{,x}; t). \tag{6.9}$$

This holds when there exists a purely inertial material inhomogeneity — i.e., inhomogeneous distribution of mass in K_R — and only a time evolution of the elasticity coefficients. This situation may be more easily realized experimentally than the general case (6.2). The time dependence in $(6.9)_2$ can only be through a relative time since the balance law has to comply with Galilean invariance. For this special case of dependency the source terms in (6.7) and (6.8) take on the following form

$$h = \frac{W}{\widehat{E}}\,\widehat{E}_t, \qquad f = \frac{K}{\hat{\rho}_0}\,\hat{\rho}_{0x}, \tag{6.10}$$

while (6.2) yields

$$u_{tt} - \hat{c}^2\, u_{xx} = 0, \qquad \hat{c}^2(x,t) = \frac{\widehat{E}(t)}{\hat{\rho}_0(x)}, \tag{6.11}$$

an equation that would lend itself to some type of space-time homogenization if periodicity is assumed in Euclidean space-time (See Section 6.7 below). However, in view of quantities that should be conserved across space-like and time-like discontinuities (see below) and the symmetry built in Equation (6.2), it might be preferable to rewrite the latter as two compatible first-order partial differential equations by introducing the auxiliary scalar field ν so that

$$\nu_t = \sigma := \widehat{E}\, u_x, \qquad \nu_x = p := \hat{\rho}_0\, u_t, \tag{6.12}$$

equations that are valid in the general case (6.2).

6.2   General properties of linear wave propagation

Considering spatial dependence of the matter density only and time dependence of the elasticity property only favours some kind of separation of space and time effects. Simple examples of smooth space-time variations of material properties have been given in Rousseau *et al.* (2011) yielding exemplary analytical solutions that may be of interest for further developments. For the time-being however, we simply note that Equation (6.11), contrary to the nonlinear case where the phase velocity would typically depend on the amplitude of the signal, shows here that the phase velocity varies from point to point in space-time while being independent of this amplitude: the theory remains *linear* although with some unusual properties.

For numerical simulations — Rousseau *et al.* (2011) —, Equation (6.11) is rewritten in the form of the hyperbolic system of the two first-order equations

$$v_t - \hat{c}^2(x, t)\, \varepsilon_x = 0, \tag{6.13}$$

$$\varepsilon_t - v_x = 0, \tag{6.14}$$

where $v = u_t$, $\varepsilon = u_x$ and $\hat{c}^2(x, t) = E(t)/\rho(x)$. This system can be solved numerically by means of the conservative wave-propagation algorithm (*cf.* Berezovski *et al.* (2008)). This numerical scheme is stable up to the value of the Courant number equal to One and is second-order accurate for smooth solutions. We refer the reader to the paper of Rousseau *et al.* (2011) for examples of stationary propagation of pulses depending on how spatial and temporal inhomogeneities vary more or less rapidly in comparison to one another.

The most remarkable acoustic-wave effect is as follows by demonstrating numerically the pure effect of time variation of the elasticity. To do this, we observe at a fixed spatial point the signal that passes by after alteration by successive increases of elasticity. A sinusoidal load is applied at the left boundary of a sample. The stress response is measured far away from that source. An *FFT* (Fast Fourier Transform) analysis of the observed signal effected by a standard MATLAB procedure shows that we obtain a shift in normalized frequency spectrum in accord with the Doppler effect formula for an observer in motion (but here it is the velocity of the signal that is observed). This differs from the more well known Doppler effect where the source is moving. Indeed, we have the standard formula

$$\frac{f}{f_0} = \frac{c(t)}{c_0} \approx 1 + \frac{\Delta c(t)}{c_0}. \tag{6.15}$$

But the case of piecewise variations considered further down in Sections 6.3 through 6.6 is of greater interest in that it allows us to critically examine what happens at junctions and accompanying jump conditions at pure space-like or pure time-like discontinuities that are typical of dynamic materials. To engage in such a study we must define purely spatial discontinuities (just like those considered in Chapter 5 that correspond to piece-wise material inhomogeneities) and time-like interface which take place at a definite time. We also have the general case of moving material interfaces that will only be briefly mentioned.

Indeed, the question naturally occurs of what happens in general not only at regular points but also at the crossing of discontinuity lines of which special cases are purely space-like or purely time-like discontinuities or interfaces. An interface is called a discontinuity surface if it has no thickness. Here, it is said to be purely space-like when material properties vary only spatially across it. Similarly, it is said to be purely time-like if the material properties vary only time wise across it. In practice a mathematically zero-thickness space-like discontinuity Σ_x often is an interface or transition layer T_x across which properties vary smoothly, albeit rapidly, in space rather than abruptly. Similarly, as it is difficult to conceive a possible instantaneous change of properties solely in time across a purely time-like discontinuity Σ_t, such a change must practically occupy a short time duration, small but not nil, during which the change occurs smoothly. This transition time τ across a thin time-like layer T_t may be that required for the switching of a rapid phase transition. Brutal spatial or temporal changes without length scale or characteristic time duration should yield the consideration of generalized functions (distributions) of the Dirac and Heaviside types. We shall avoid this mathematical complication. But there remains the question of what are the conditions imposed on the field quantities (here, those derived from the particle motion) across Σ_x or Σ_t because the situation is quite different for an interface Σ_x and a time line Σ_t.

If we examine the question for piecewise constant variations in space and time with 1D space dimension and a space-time diagram, then discontinuities Σ_x are represented by straight lines parallel to the t-axis while discontinuities Σ_t are straight lines parallel to the x-axis (see Figure 6.1). Of course the 1D spatial situation here also may be misleading because what happens in space is essentially typically multidimensional with a *co-vectorial* connotation. In effect the dual of position is the canonical momentum according to the phase definition in Chapter 4 and in terms of the wave vector a vector direction is involved with possible change of orientation across a

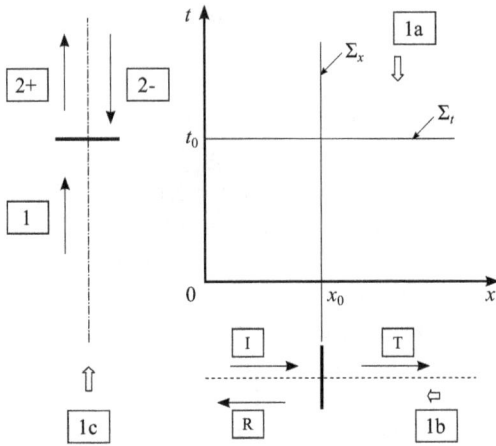

Fig. 6.1   Space-like and time-like discontinuities.

two-dimensional interface surface Σ_x — e.g., in Snell–Descartes law — or across T_x). Quite differently, it is always a *scalar* quantity, *energy* or *frequency* as shown by the duality present in the definition of the phase, which is the essential evolving quantity at a time-like interface Σ_t or across T_t. We shall first examine these two cases from the point of view of the wave picture and then pay special attention to the quasi-particle re-interpretation for a time-like interface which is the most original case.

6.3   Case of a fixed material interface or transition layer

This is represented by a point at x_0 in Figure 6.1 (Part 1b). There is no matter transfer at that point although both density and elasticity may change abruptly across Σ_x. There is continuity of the normal surface traction. Both **P** and **K** are not kept unchanged at Σ_x. The only quantity that evolves is the gradient of displacement, as frequency is kept fixed and wave vector evolves (in 2D we obtain the Snell–Descartes law). For a transition layer T_x where only the density ρ varies and does it slowly enough, application of the WKB method yields with a constant frequency ω a variable wave number $k(x) = \omega/c(x)$ with, in 2D, a curvature of rays — described by a relation of the form $\sin\theta(x) = \big(\sin\theta(0)/c(0)\big)\,c(x)$ — and a *concentration of energy*. If both density and elasticity vary across T_x, then we obtain an inhomogeneous wave equation, a case not envisaged here. With a normal incidence we are in the situation studied in Chapter 5.

For the discontinuity Σ_x we must check at, say $x = x_0 = 0$, and *for all values* of time t, the continuity conditions (displacement and stress)

$$u_1 = u_2, \qquad u_{1x} = u_{2x}, \tag{6.16}$$

with $u_1 = u_I + u_R$ and $u_2 = u_T$ (Figure 6.1, Part 1b) since in general there is one incoming wave (on side 1), one transmitted wave (on side 2) and a reflected wave (on side 1). Then we deduce immediately that, since $E_1 = E_2$ and $\rho_1 \neq \rho_2$, then $c_1 \neq c_2$, and

$$\omega_1 = \omega_2 = \omega, \qquad k_1 \equiv \frac{\omega_1}{c_1} \neq k_2 \equiv \frac{\omega_2}{c_2}. \tag{6.17}$$

6.4 Case of a time-line or thin time-like interface layer

In this case we consider only the possibility of an evolving elasticity (an evolving density would yield not only no mass conservation but also an inhomogeneous wave equation). This is represented by a point at t_0 in Figure 6.1 (Part 1c). But here, analogous to the mass-flux condition and continuity of normal traction at Σ_x, we have continuity of the displacement gradient and of the displacement itself, respectively. The wave vector does not evolve but the velocity varies. For a "transition layer" T_t with a typical time scale τ much larger than the acoustic period, the equivalent of the WKB method in time yields $\omega(t) = k\,c(t)$ with k fixed and thus

$$\omega(t) = \omega_0 \frac{c(t)}{c(0)}, \tag{6.18}$$

hence *a Doppler effect with capture of energy*. That is what distinguishes the two types of transitions. As a partial conclusion, spatial inhomogeneity allows the convergence or divergence of wave by conservation of the momentum, while a dynamic medium with time inhomogeneity allows capture of energy from the outside with a resulting change in frequency. The combination of the two in a true dynamic material may yield a concentration of energy although the system is fully linear, as shown by Lurie (2007).

Because of the built-in symmetry between spatial and temporal behaviours, we will know what to do at Σ_t once we know the conditions at Σ_x. In the latter case for a fixed Σ_x we have continuity of the mechanical traction, i.e., in terms of jumps,

$$[\widehat{E}\,u_x] = 0 \quad \text{at } \Sigma_x, \tag{6.19}$$

hence this translates to

$$[\hat{\rho}_0 \, u_t] = 0 \quad \text{at } \Sigma_t. \tag{6.20}$$

In the considered special case, these two conditions reduce to

$$[u_x] = 0 \text{ at } \Sigma_x, \qquad [u_t] = 0 \text{ at } \Sigma_t, \tag{6.21}$$

where square brackets denote the usual jump.

For a time-like interface Σ_t situated, say, at $t = t_0 = 0$, we must check, *for all values of x*, [compare to (6.16); $\rho_1 = \rho_2$, $E_1 \neq E_2$] that

$$u_1 = u_2, \qquad u_{1t} = u_{2t}, \tag{6.22}$$

where region 1 is below Σ_t in space-time and region 2 is above. Here we should pay attention to the wording because we are no longer speaking of propagating waves since the transition across Σ_t occurs in the ideal case parodying the case (6.16) at zero propagation velocity (vertical lines crossing Σ_t). Nonetheless, a solution will exit at Σ_t (*cf.* Lurie (2007); Ginzburg and Tsytovich (1979)) only if we have one "signal" A_1 coming from region 1 toward region 2, one signal A_2^+ continuing in region 2 and, what is more surprising, one signal A_2^- in region 2, oriented from region 2 to region 1 (*as if it were coming from the future*). That is, we have (*cf.* Figure 6.1, Part c): u_1 and $u_2 = u_+ + u_-$. A simple estimate yields (with $\rho_1 = \rho_2$, $E_1 \neq E_2$, $c_1 \neq c_2$)

$$T_{12}^+ = \frac{A_2^+}{A_1} = \frac{c_2 + c_1}{2\,c_2}, \qquad T_{12}^- = \frac{A_2^-}{A_1} = \frac{c_2 - c_1}{2\,c_2}. \tag{6.23}$$

Then at Σ_t, we have (since $c_1 \neq c_2$)

$$k_1 \equiv \frac{\omega_1}{c_1} = k = k_2 \equiv \frac{\omega_2}{c_2}, \qquad \omega_1 \neq \omega_2. \tag{6.24}$$

6.5 Quasi-particle re-interpretation at a time-like interface layer

It is clear that a signal coming from the future as hypothesized above is *not* physical. The introduced image is just an artefact coming from an unduly applied analogy with what happens at a space-like interface of material inhomogeneity. To proceed we consider the interaction of an incoming 1D linear elastic wave $u_0(x, t) = U \cos(k\,x - \omega\,t)$ in a region of density, elasticity coefficient and elastic speed ρ_0, E_0, $c_0 = \sqrt{E_0/\rho_0}$, with a time-like discontinuity surface Σ_t at time $t = \tau$ to a region of Euclidean space-time with density, elasticity coefficient and elastic speed ρ_0, E_1, $c_1 = \sqrt{E_1/\rho_0}$. According to Eq. (6.22) above we have the condition (no change in wave number)

$$k = \frac{\omega_0}{c_0} = \frac{\omega_1}{c_1}. \tag{6.25}$$

The change in elasticity coefficient is due to an instantaneous *input of energy* of amount $S_{0,1}$ at Σ_t. The system acts like a "pump" (as called in laser technology but here for elastic waves). As a matter of fact, exploiting the vision of elastic waves as "quasi-particles" like in the transmission problem examined in Chapter 5, and computing the changes in both kinetic and deformation energies in terms of the quasi-particle properties, we prove that

$$E_1^{\text{cin}} - E_0^{\text{cin}} = \frac{1}{2} \rho_0 \, k \, \pi \, U^2 \left(c_1^2 - c_0^2 \right) = \frac{1}{2} S_{0,1}, \tag{6.26}$$

and

$$E_1^{\text{def}} - E_0^{\text{def}} = \frac{1}{2} k \, \pi \, U^2 \left(E_1 - E_0 \right) = \frac{1}{2} S_{0,1}. \tag{6.27}$$

This shows that there is *equipartition* of the input energy in this linear elastic system, half of it going to the change of kinetic energy and the other half to the change in deformation energy.

The "mass" of the quasi-particle impinging on Σ_t and that of the transmitted "quasi-particle" are identical and read

$$M_0 = M_1 = \rho_0 \, k \, \pi \, U^2. \tag{6.28}$$

If $S_{0,1}$ is known then the change in the squared velocity of the quasi-particle through Σ_t is given by the simple formula

$$c_0^2 - c_1^2 = \frac{S_{0,1}}{M_0}. \tag{6.29}$$

In this vision the reasoning is much simpler with quasi-particles than with waves. In particular, this replaces the specious argument considering the displacement fields u_0 and $u_1 = u_+ + u_-$ of the wave interpretation. This, in turn, allows one to examine more complex cases such as a succession of close time-like discontinuities or, more realistically, a time interval through which the input of energy is continuous.

In the first case, imagine there is a sequence of n time-like discontinuities. With the same labelling system as above we can write

$$\frac{1}{2} S_{n-1,n} = E_n^{\text{cin}} - E_{n-1}^{\text{cin}} = \frac{1}{2} \rho_0 \, k \, \pi \, U^2 \left(c_n^2 - c_{n-1}^2 \right), \tag{6.30}$$

and

$$\frac{1}{2} S_{n-1,n} = E_n^{\text{def}} - E_{n-1}^{\text{def}} = \frac{1}{2} k \pi U^2 \left(E_n - E_{n-1} \right). \tag{6.31}$$

Just like before we have

$$M_n = M_{n-1} = \ldots = M_0, \tag{6.32}$$

and for the whole sequence of time-like transitions

$$\frac{1}{2} \sum_{\alpha=0}^{\alpha=n} S_{\alpha-1,\alpha} = E_n^{\text{cin}} - E_0^{\text{cin}} = \frac{1}{2} \rho_0 k \pi U^2 \left(c_n^2 - c_0^2 \right), \tag{6.33}$$

and

$$\frac{1}{2} \sum_{\alpha=0}^{\alpha=n} S_{\alpha-1,\alpha} = E_n^{\text{def}} - E_0^{\text{def}} = \frac{1}{2} k \pi U^2 \left(E_n - E_0 \right). \tag{6.34}$$

Now we can pass to a continuous variation of the energy input. We clearly have

$$M_0 = \rho_0 k \pi U^2, \tag{6.35}$$

$$E(t) = E_0 + \int_0^t \frac{f(\tau)}{(M_0/\rho_0)} d\tau, \tag{6.36}$$

$$S(t) = \int_0^t f(\tau) d\tau. \tag{6.37}$$

In this thought experiment the elastic medium has a time-varying elasticity, and the associated quasi-particle has a constant mass but a varying velocity.

Trajectory through a time-like surface followed by a space-like abrupt inhomogeneity

We refer to Figure 6.2 below in which a linear acoustic signal first travels in a region indexed 0, then interacts with a time-like interface Σ_t of vanishing thickness acquiring an energy S_1 (at point A), travelling then in a region (indexed 1) of altered elasticity coefficient with a larger speed than before, and then crossing a space-like interface Σ_x at point B with a change in matter density (material inhomogeneity in the inertial property) and then travelling in region indexed 2. As seen above in the quasi-particle interpretation, before point A, the mass of the quasi-particle is M_0, with wave number k_0 and characteristic speed $c_0 = \sqrt{E_0/\rho_0}$. At point A the input S_1 of energy is shared equally between the kinetic and deformation energies of the wave, keeping the same wave number, $k_1 = k_0$, and the same quasi-particle mass,

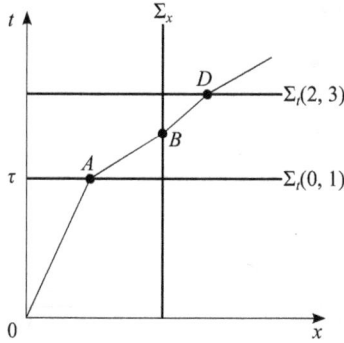

Fig. 6.2 Trajectory of a ray in a dynamic material.

$M_1 = M_0$, but transforming the frequency ω_0 into ω_1 and the speed into $c_1 = \sqrt{E_1/\rho_0}$. At point B the frequency is left unchanged, $\omega_2 = \omega_1$, but the incident wave mode gives rise to a transmitted signal with index $2 = T$ and a reflected one indexed R, with $k_T = \omega_2/c_2 = \omega_1/c_1$ (with $c_2 < c_1$ if $\rho_2 > \rho_1$) and $k_R = k_1 = k_0$ with respective quasi-particle masses M_T and M_R (see Chapter 5). The transmitted signal has amplitude reduced to TU with a transmission coefficient T, and therefore a "mass" M_T proportional to $T^2 U^2$ with velocity c_2. The reflected signal progresses with increasing time but comes back towards the initial source with an amplitude RU and reflection coefficient R, and therefore a mass M_R proportional to $R^2 U^2$ and velocity c_1. We can imagine that the forward travelling signal in region 2 meets again — at point D — a new time-like interface Σ_t where it receives a new energy input S_2 with invariant wave number $k = k_T$ and a conserved mass $M = M_T$, but a new alteration in frequency yielding $\omega = \omega_3$ and so on. We see that the reasoning exploiting the re-interpretation in terms of quasi-particles is much simpler than the one limited to the wave picture. This naturally leads us to consider waves travelling back and forth in a rod of finite length and their associated quasi-particles.

6.6   Waves along a rod of finite length

We consider the elastic wave motion back and forth along a rod of finite length a (See Figure 6.3) and the following scenario. There is a vacuum at both ends of the rod so that we observe only internal reflections at both ends since the vacuum does not support elastic waves. At time $t = 0$ a

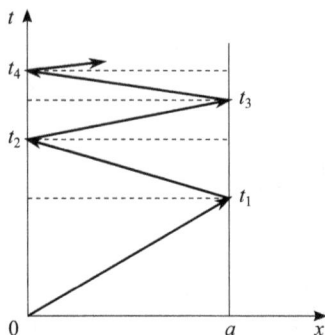

Fig. 6.3   Wave travelling forth and back in a rod of finite length with energy input at each end.

signal of frequency ω_0 and amplitude U is emitted at the origin $x = 0$. It travels down the rod at speed c_0 until it reaches the end point $x = a$ at time t_1. At that very moment (point 1 in Figure 6.3), due to some mechanism, it captures energy S_1 from a time-like line which we assume transforms the speed into a double value $2\,c_0$. The so reflected signal now travels back to the origin which it reaches at time $t_2 = t_1 + t_1/2$ (point 2). At this point it captures energy S_2 — at another time-like line — causing an increase of speed to $(3\,c_0)$. The emerging signal now travels in the positive direction and reaches point 3 at $x = a$ at time $t_3 = t_1 + t_1/2 + t_1/3$. A new capture of energy S_3 is caused at this point and governs the new signal that goes to the left, *etc.* The imposed inputs of energy are given by formulas of the type of (6.26) or (6.27).

In summary, we have:

- at point 0: $t = 0$, $x = 0$;
- at point 1: $t = t_1 = c_0/a$, $x = a$, and
$$S_1 = \left((2\,c_0)^2 - c_0^2\right) M_0^2 = 3\,M_0^2\,c_0^2;$$
- at point 2: $t = t_2 = t_1\,(1 + 1/2)$, $x = 0$, and
$$S_2 = \left(3^2 - 2^2\right) M_0^2\,c_0^2 = 5\,M_0^2\,c_0^2;$$
- at point 3: $t = t_3 = t_1\,(1 + 1/2 + 1/3)$, $x = a$, and
$$S_3 = 7\,M_0^2\,c_0^2;$$

and continuing this scenario:

- at point $n - 1$: $t = t_{n-1} = t_1\left(1 + 1/2 + 1/3 + \ldots + 1/(n-1)\right)$, $x = a$, and
$$S_{n-1} = \left(n^2 - (n-1)^2\right) M_0^2\,c_0^2;$$

- at point n: $t = t_n = t_1 (1 + 1/2 + 1/3 + \ldots + 1/n) = t_1 \sum_{k=1}^{n} 1/k$, $x = 0$, and

$$S_n = ((n+1)^2 - n^2) M_0^2 c_0^2 = (2n+1) M_0^2 c_0^2.$$

In these expressions $M_0 = \rho_0 \omega_0 \pi U^2 / c_0$ where ω_0 and U are the angular frequency and magnitude of the initial signal at $(t = 0, x = 0)$ — see Equation (6.35). We note that $(t_n - t_{n-1}) \to 0$ as n increases.

Each time some energy is put in, the elastic medium becomes more rigid, and the speed of propagation is increased, so that the quasi-particle associated with the wave process travels faster and faster between the two ends of the rod. In terms of wave properties, both circular frequency and speed are increased at each input of energy but the wave number is always left unchanged since the signal does not cross a material inhomogeneity. If we continued the procedure ad infinitum, the rod would become completely rigid and the propagation would become instantaneous, a case obviously easily grasped in the quasi-particle (point-mechanics) picture. This somewhat academic example illustrates well the difference in interpretation between the wave picture and the associated quasi-particle re-interpretation.

6.7 Space-time homogenization of dynamic materials

As shown in foregoing sections, the canonical equations of (non)-conservation of energy and wave momentum (6.7) and (6.8) play a fundamental role in associating quasi-particles with wave modes propagating in dynamic materials. We recall the reader that these two equations, with nonvanishing source terms in their right-hand side, reflect the lack of homogeneity in time and space, respectively, of the material. Authors concerned with spatial homogenization (e.g., Sanchez–Palencia and Zaoui (1987)) seldom — or never — refer to the canonical conservation laws of the studied system although the conservation — or non-conservation — of field or wave momentum is the best evidence for homogeneity or the lack of homogeneity in space. With dynamic materials under consideration, a naturally raised question is what are the circumstances in which the material may be looked upon as a homogeneous material, in both time and space? In other words, is it possible to proceed to a mathematical homogenization of such systems?

Of course, space-time homogenization is a seldom considered field. However, among other questions of control theory, Lurie (2007) has reported the homogenization of system (6.11) or (6.12) considering some asymptotic

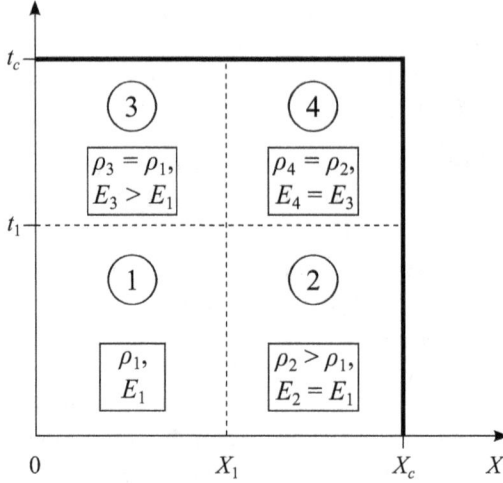

Fig. 6.4   Dynamic materials: checkerboard configuration.

expansion of the u solution for long times compared to the period of the repeated motif (e.g., a checkerboard such as in Figure 6.4) in Euclidean 2D space-time (*cf.* Lurie and Weekes (2006); Lurie *et al.* (2009)). To examine this matter we first point out two special cases of interest.

6.7.1   *So-called "slow" time-like configuration*

With dynamical processes in view it is natural to look for solutions that depend on time and a travelling coordinate, i.e. the new variables

$$\zeta = x - ct, \qquad \tau = t. \tag{6.38}$$

For comparatively slow variations of the material coefficients in time, these coefficients are taken function of the fast variable

$$\xi = \varepsilon^{-1}\zeta, \tag{6.39}$$

where ε is an infinitesimally small parameter. With $F(x,t)$ replaced by $\widetilde{F}(\zeta,t)$, the basic Lagrangian density (6.1) takes on the form

$$\widetilde{L} = \frac{1}{2}\,\tilde{\rho}_0\left(\frac{\partial\tilde{u}}{\partial\tau}\right)^2 - \tilde{\rho}_0\,c\,\frac{\partial\tilde{u}}{\partial\tau}\,\frac{\partial\tilde{u}}{\partial\zeta} + \frac{1}{2}\left(\tilde{\rho}_0\,c^2 - \widetilde{E}\right)\left(\frac{\partial\tilde{u}}{\partial\zeta}\right)^2. \tag{6.40}$$

On applying the canonical formalism recalled Chapter 3 and remembering (6.39) and accounting for the Lagrangian (6.40), we see that the source term in the "energy" equation (6.7) vanishes at the present order of approximation. Thus, *the system appears as "homogeneous in time"*, exhibiting neither expense nor capture of energy. This is noticed by Lurie (2009, p. 338). The material remains only materially inhomogeneous with possible discontinuities of the Σ_x type only. Note also, following Lurie, that dynamical homogenization is feasible only if the wave speed c is such that $c^2 < \tilde{c}_0^2 := \tilde{E}/\tilde{\rho}_0$ in order that the energy density (Hamiltonian density) $H = \tilde{u}_\tau \tilde{p} - \tilde{L}$ (Legendre transformation) remains positive. It can be said that we have a *slow* configuration.

6.7.2 So-called "fast" space-like configuration

In strict parallel to the reasoning in the preceding paragraph, we now look for dynamical solutions that depend on space and time through the new space-time variables

$$\zeta = x, \qquad \theta = t - x/c. \tag{6.41}$$

With $F(x,t)$ replaced by $\overline{F}(\zeta,\theta)$, the basic Lagrangian density (6.1) takes on the form

$$\overline{L} = \frac{1}{2}\left(\bar{\rho}_0 - c^{-2}\,\overline{E}\right)\left(\frac{\partial\bar{u}}{\partial\theta}\right)^2 + c^{-1}\,\overline{E}\,\frac{\partial\bar{u}}{\partial\theta}\,\frac{\partial\bar{u}}{\partial\zeta} - \frac{1}{2}\,\overline{E}\left(\frac{\partial\bar{u}}{\partial\zeta}\right)^2. \tag{6.42}$$

On applying again the canonical formalism with independent variables (6.41), we see that the source term in the "wave momentum" equation (6.8) vanishes at the considered order of approximation. *The system then appears as "materially homogeneous"* with vanishing force of inhomogeneity. Its possible inhomogeneity is only in time with possible discontinuities of the Σ_t type only. For the feasibility of the homogenization procedure it is checked that the wave speed c must be such that $c^2 > \bar{c}^2 := \overline{E}/\bar{\rho}_0$ in order that the corresponding energy (Hamiltonian) density $H = \bar{u}_\theta\,\bar{p} - \overline{L}$ remains positive. We can say that we have a *fast* configuration.

The results of these two paragraphs show the technical possibility of homogenization with respect to the two variables (6.38)$_1$ and (6.41)$_2$. To do this, however, we need to complement the asymptotics with an averaging procedure of some kind.

6.7.3   *Space-time homogenization for a long time*

Considering independent variables such as in (6.38) we look at asymptotic expansions in the form

$$F = \widehat{F}(x, t = \tau, \xi) = F_0(x, \tau, \xi) + \varepsilon\, F_1(x, \tau, \xi) + \varepsilon^2\, F_2 + \ldots, \qquad (6.43)$$

where we have introduced the *fast* space variable (6.39). But now we assume that the material parameters are periodic functions of period one in this fast variable, so that we can envisage a periodic homogenization. This matter was addressed by Lurie (in particular, 2007, Sections 2.2–2.5), for which the system (6.11) can be rewritten as the system

$$\bar{\rho}_0\, u_\tau = \bar{\rho}_0\, c\, u_\zeta + v_\zeta, \qquad v_\tau = c\, v_\zeta + \overline{E}\, u_\zeta, \qquad (6.44)$$

in terms of the pair of variables (6.12). From this one solves for the space-like derivatives as

$$u_\zeta = (c/\Delta)\, u_\tau - (\bar{\rho}_0\, \Delta)^{-1}\, v_\tau, \qquad v_\zeta = -(\overline{E}/\Delta)\, u_\tau + (c/\Delta)\, v_\tau, \quad (6.45)$$

with

$$\Delta := c^2 - \tilde{c}^2, \qquad \tilde{c}^2 := \overline{E}/\bar{\rho}_0. \qquad (6.46)$$

Equations (6.45) are in a convenient form for easy averaging because the field derivative functions in these equations are unaffected by an averaging that involves only the material functions that are supposed to be periodic in the fast variable (6.39). Here, for the sake of simplicity, the material is viewed as made of two constituents $\alpha = 1, 2$ with respective volume fractions m_α, somewhat like the spatial arrangement of the two bottom sub-cells in Figure 6.4. Denoting by the symbol $\langle\,.\,\rangle$ the arithmetic average, we write

$$B = \langle \Delta^{-1} \rangle, \qquad C = \langle (\bar{\rho}_0\, \Delta)^{-1} \rangle, \qquad D = \langle \overline{E}/\Delta \rangle, \qquad (6.47)$$

so that the average of equations (6.45) reads

$$u_\zeta = c\, B\, u_\tau - C\, v_\tau, \qquad v_\zeta = -D\, u_\tau + c\, B\, v_\tau. \qquad (6.48)$$

From this one may return to equations governing v_t and v_x (*cf.* equations (6.11)), but now with coefficients involving the averaged quantities, e.g., formally

$$v_t = p\, u_x - q\, u_t, \qquad v_x = q\, u_x + r\, u_t. \qquad (6.49)$$

We refer to Lurie (2007, pp. 24–28), for the expression of the coefficients p, q, and r in general and special cases of configurations. This author shows on account of the asymptotic solution (6.43) that the initial wave equation renders, for the u_0 function, the equation

$$r\,u_{0tt} + 2\,q\,u_{0xt} - p\,u_{0xx} = 0. \tag{6.50}$$

It is readily shown that this equation is derivable for the following peculiar Lagrangian density:

$$L_0 = \frac{1}{2}\left(r\,u_{0t}^2 - p\,u_{0x}^2 + 2\,q\,u_{0t}\,u_{0x}\right). \tag{6.51}$$

Accordingly, on using the canonical formalism, for the homogenized zeroth-order solution, we obtain the conservation equations of energy and canonical momentum in the following *strict conservative* form:

$$\frac{\partial H_0}{\partial t} - \frac{\partial Q_0}{\partial x} = 0, \qquad \frac{\partial p_0^w}{\partial t} - \frac{\partial b_0}{\partial x} = 0. \tag{6.52}$$

In other words, at this order of approximation, after a long time — or equivalently a long distance — of propagation the system appears to be *homogeneous both in time and space*. Of course the crucial step here — due to Lurie and not repeated — was obtaining the reduced equation (6.50). The result is tantamount to saying that the looked for effects disappear altogether by successive increases and decreases that compensate each other since the inequalities requested in Paragraphs 6.7.2 and 6.7.3 above must be satisfied In general, however, in order to realize practically the phenomenon, it remains to find a way to cause a sufficiently rapid and sizeable change in time of the elasticity properties by action of an external field causing, e.g., a fast phase transition.

6.8   Generalization to moving interfaces

In the above given developments we have considered fixed material disconti-nuities in space and time-like lines at specified times. A much more general problem would consist in considering a *moving material interface* $\Sigma_{x,t}$. The latter may represent a moving transition front. In this case, although the problem remains one-dimensional in space (and therefore in fact the dis-continuity surface is still reduced to a point along the x-axis), the *spatial interface* moves in time in the material. The natural approach then is to consider the wave problem as seen by an observer moving with the ma-terial interface, i.e., to consider an instantaneous Galilean transformation such that the new space-time coordinates be

$$\xi = x - \hat{x}(t), \qquad \tau = t, \tag{6.53}$$

where the interface instantaneous position is given by $\hat{x}(t)$. With constant velocity V starting from $x = 0$ we have $\hat{x}(t) = V\,t$. This velocity must be compared to the characteristic velocity of the material on its two faces (in principle at each instant of time) with the constraint that the considered mathematical system be *hyperbolic*. This matter was examined by Lurie (2007, Chapter 2) — who concludes that a necessary condition for the existence of a required solution at $\Sigma_{x,t}$ is that

$$\left(V^2 - c_1^2\right)\left(V^2 - c_2^2\right) > 0, \tag{6.54}$$

where c_1 and c_2 are the characteristic phase velocities on the two sides of the interface. Note that moving interfaces in 1D corresponding to phase changes have been dealt analytically by Ericksen (1991) while 2D ones have been numerically simulated by Berezovski and Maugin (see Berezovski *et al.* (2008); and in Maugin (2011b, Chapter 13), with coloured pictures). But these works did not envisage the superimposition of harmonic signals as done in the present chapter for "dynamic" materials.

6.9   Conclusion

As the main conclusion to this chapter, we first note that the simultaneous space and time inhomogeneities that are characteristic of dynamic materials, naturally called for the consideration of canonical conservation laws of energy and wave momentum. In turn, this directly led to the introduction of the notion of quasi-particles. The second point relates to the inherent originalities in this framework, that is, the occurrence of space-like and time-like discontinuity surfaces in space-time. This is where the comparison between the wave-like picture and the associated quasi-particle one is the most enlightening. For a *geometric interface* (material discontinuity) in the first vision, we classically are in presence of an incident wave, a reflected wave, and a transmitted one. In the quasi-particle vision we correspondingly associate incident, reflected, and transmitted particles with respective masses defined in terms of the properties of the medium before and after the discontinuity. We admit that both approaches are here essentially equivalent. The situation is altogether different for a *time-like interface*. As shown in Sections 6.5 though 6.7, while the "wave" picture becomes untenable — because of the artificiality of introducing a "signal" coming from the future —, the quasi-particle picture is much more conceptually pleasant with masses defined before and after the time line, with a change in speed (related to an evolution in the elasticity coefficient) because of the input of energy. Note in conclusion that the inverse process — losing energy to the exterior — is in principle possible, but poses a problem for its experimental realization (this is *not* viscosity; this extraction of energy must be caused by some external field that would suck out energy from the elastic wave impinging on a time-like line).

Chapter 7

Elastic surface waves in terms of quasi-particles

Object of the Chapter

Surface elastic waves are of mixed theoretical, (mechanical) experimental, and geophysical interest. They concern signals that have their energy confined to the vicinity of a limiting surface. Translated in the interpretation of quasi-particles they offer good examples of guided moving quasi-particles. This is established here for the case of Rayleigh surface waves, the same but perturbed by a surface energy, the pathological case of leaky surface waves, the case of Love waves, and that of shear waves guided by a surface endowed with both elastic energy and inertia (mass) in agreement with a concept due to Tiersten and Murdoch. With the exception of the leaky waves, the other waves, whether non-dispersive or dispersive, all have a point thermomechanics associated with quasi-particles in steady Newtonian–Leibnizian motion in the absence of dissipation, sometimes with a relativistic connotation.

7.1 The notion of surface wave

Elastic surface waves or surface acoustic waves (SAWs) provide a beautiful and useful example of dynamical problems in elasticity. They correspond to elastic waves that are guided by the surface of a semi-infinite space. Because of the prevailing role of the guiding surface the importance of boundary conditions cannot be overlooked. While field equations provide rather general solutions in the bulk of the body, these boundary conditions provide the final solution of the propagation problem in the form of the *dispersion relation*, i.e., the necessary relation between frequency and wave number. The usefulness of the concept of elastic surface waves was first recognized in geophysics. But the practical interest for such waves was en-

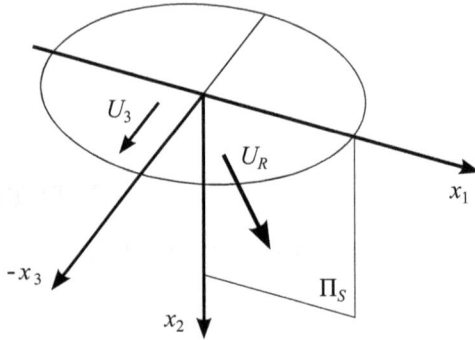

Fig. 7.1   Surface wave problem: $\mathbf{U}_R$: Rayleigh vector component; U_3: SH component (x_1: direction of propagation; material body: $x_2 > 0$).

hanced by their applications in signal processing, nondestructive evaluation techniques, and the propagation of dislocations and cracks.

The sketchy Figure 7.1 allows us to introduce some basic vocabulary and to define more precisely the different types of surface waves. The following notation is implemented. The elastic material half space corresponds to the part $x_2 > 0$ of Euclidean physical space. The direction of propagation of the wave mode is along the axis x_1. The rectangular coordinate system is completed by x_3. The limiting plane is none other than the plane $x_2 = 0$. Planes parallel to the basic plane (x_1, x_2) are called *sagittal* planes and are denoted by Π_S. The solution is considered invariant by translation along the transverse coordinate. For the semi-half space $x_2 < 0$, we usually have a vacuum, and occasionally another material such as a fluid or another elastic material. The most familiar SAWs are:

- the **Rayleigh SAW** (after Lord Rayleigh (1887): This is a wave with displacement parallel to Π_S made of a P (for "pressure") component parallel to the direction of propagation and an SV (for "shear vertical") component along x_2, that propagates along the top free surface of a linear elastic isotropic homogeneous half space;
- the **Love SAW** (after A. E. H. Love (1911)): This is a wave with pure SH (for "shear horizontal") component that is guided along a structure made of a layer of "slow" elastic material superimposed on an elastic half space. This wave mode is particularly relevant to geophysical studies and damage caused by earthquakes;
- the **Stoneley SAW** (after R. Stoneley (1924)): This is a wave

that is guided by the interface between two elastic half spaces of appropriate relative properties;

- the **Bleustein–Gulyaev SAW** (after Bleustein (1968), and Gulyaev (1969)): This is an electroelastic wave of mechanical SH polarization whose existence is due to a surface coupling between stress and electric field.

To these we add for further consideration

- the **Murdoch SAW** (after Murdoch (1976)): This is an elastic wave that is guided by the limiting top surface which itself has inertial and elastic properties.

Most of these surface wave modes are well documented in classical books on elastic wave propagation (e.g., Achenbach (1973); Auld (1973); Dieulesaint and Royer (2000); Brekhovskikh (1960)) to which we refer the reader for details. There are more general cases such as SAWs propagating on top of an anisotropic elastic half-space, on a magnetostrictive half-space, on a ferromagnetic half space coupling elastic and spin waves, on semi-conductors, and also numerous cases of perturbations of the limiting surface: curvature, roughness, gratings, elastic energy, waves in elastic layers (Lamb), *etc.* Only the simplest standard cases will be considered hereinafter.

The matter of surface acoustic waves has also been considered from the point of view of solid state physics. This led to the introduction of the notion of *surface acoustic phonon* where the definition of such quasi-particles relies essentially on a lattice-dynamics approach (Ludwig (1991); Henrich and Cox (1994)). The interest in such studies stems from the potentially associated simple interpretation of the interaction between fellow waves or of the interaction of such a wave with material objects (defects, inclusion,...). But the present Chapter aims at something else: to associate quasi-particles to *surface waves or guided waves* that are *known* solutions of elastic wave problems associated with specific boundary conditions such as Rayleigh waves, *etc.* Remember that such waves have an energy confined to the vicinity of the appropriate guiding interface; and their amplitude decreases with depth in the substrate. We are thus led to a clear introduction of the notion of *surface acoustic phonon* but basing on the continuous framework. Perturbations of various kinds (surface energy, connection to an external fluid, nonlinearity, viscosity of the substrate) are also envisaged. The technique employed for this study is based on the exploitation of the conservation equations for energy and wave-momentum deduced via

Noether's theorem or by direct computation, and associated with the basic field equations of which we already know — sometimes painstakingly — the analytic continuum solutions. Volume integration of the conservation laws is then performed over a volume representative of the surface wave motion (See coloured Plate 7.1). This accounts for the wavelength of the wave and the depth behaviour. Only pure linear elasticity for the substrate is considered in this Chapter, electroelastic couplings being reserved to Chapter 8. The continuum solutions will be recalled without much detail, referring the reader to the already mentioned bibliography on SAWs.

Here we focus first on the case of the Rayleigh surface wave of which the classical solution is recalled in Section 7.2. The associated conservation of wave momentum is then given with the accompanying resulting "grain of SAW", the looked for quasi-particle. The latter behaves like a Newtonian mass point in inertial motion with a mass that depends on the whole of the continuum solution including on the properties of amplitude decrease in the depth of the substrate. The influence of a possible surface energy (but no inertia/mass) is also examined, while that of a coupling with an external inviscid fluid is the subject of another paragraph. This yields the notion of leaky surface waves in the continuum description and strange particle properties in the quasi-particle re-interpretation. The case of Love wave is dealt with in Section 7.3. The case of Murdoch waves, also purely mechanical, is the subject of Section 7.4. The effects of viscosity and nonlinearity in the substrate are delayed until Chapter 8 where the involved elastic displacement is a scalar only (SH wave!), and therefore lends itself to an easier full analytic treatment. Much of the presented material is borrowed from three papers of the authors (Rousseau and Maugin (2011a, 2012b, 2014)) and the synthesis by Maugin and Rousseau (2013).

7.2   The Rayleigh surface wave in isotropic linear elasticity

7.2.1   *Definition of Rayleigh waves*

We consider a linear elastic semi-infinite isotropic homogeneous medium limited by the plane $x_2 = 0$. Its mechanical constitutive equation is the well known Hooke law given in Cartesian tensorial form by

$$\sigma_{ij} = \lambda \varepsilon_{kk} \delta_{ij} + 2 \mu \varepsilon_{ij}, \qquad \varepsilon_{ij} = \frac{1}{2} (u_{i,j} + u_{j,i}), \qquad (7.1)$$

where λ (not to be confused with a wavelength) and μ are the Lamé coefficients. The field equations and boundary conditions are (no body force, no traction applied at the surface)

$$\frac{\partial}{\partial x_i}\sigma_{ij} = \rho_0 \frac{\partial^2 u_j}{\partial t^2} \quad \text{for } x_2 > 0, \tag{7.2}$$

$$n_i\,\sigma_{ij} = 0 \quad \text{at } x_2 = 0. \tag{7.3}$$

This purely mechanical case is of interest because, being the simplest one in its class, it singles out an elastic displacement with two components along x_1 and x_2-axes, hence possesses an elastic displacement vector polarized parallel to Π_S. Indeed, the Helmholtz decomposition of a vector field allows us to write

$$\mathbf{u} = \mathbf{u}_L + \mathbf{u}_T \quad \text{with } \nabla \cdot \mathbf{u}_L = 0, \quad \nabla \times \mathbf{u}_T = \mathbf{0}. \tag{7.4}$$

Then $\mathbf{u}_L$ and $\mathbf{u}_T$ satisfy wave equations:

$$\nabla^2 \mathbf{u}_L - c_L^{-2}\frac{\partial^2}{\partial t^2}\mathbf{u}_L = \mathbf{0}, \quad \nabla^2 \mathbf{u}_T - c_T^{-2}\frac{\partial^2}{\partial t^2}\mathbf{u}_T = \mathbf{0}, \tag{7.5}$$

where

$$c_L = \sqrt{\frac{\lambda + 2\mu}{\rho_0}}, \quad c_T = \sqrt{\frac{\mu}{\rho_0}}, \tag{7.6}$$

are respectively the speeds of longitudinal and transverse waves.

In the case of plane strains in the (x_1, x_2)-plane and if we assume that the fields $\mathbf{u}_L$ and $\mathbf{u}_T$ are two harmonic plane waves, we have

$$\begin{aligned}
\mathbf{u}_L &= \Re\{\mathbf{a}_L \exp\left(i\left(k_1\,x_1 + k_{2L}\,x_2 - \omega\,t\right)\right)\}, \\
\mathbf{u}_T &= \Re\{\mathbf{a}_L \exp\left(i\left(k_1\,x_1 + k_{2T}\,x_2 - \omega\,t\right)\right)\},
\end{aligned} \tag{7.7}$$

with wave numbers (ω is the angular frequency)

$$k_L = \frac{\omega}{c_L}, \quad k_{2L} = (k_L^2 - k_1^2)^{1/2}, \quad k_{2T} = (k_T^2 - k_1^2)^{1/2}, \quad k_T = \frac{\omega}{c_T}. \tag{7.8}$$

Then the Rayleigh wave is an *acoustic eigen-mode* corresponding to a particular coupling between longitudinal (P) and transverse (SV) waves. This mode is a surface wave propagating at the *free* surface of an elastic solid semi-infinite space and polarized in the sagittal plane. The boundary conditions

$$\sigma_{12} = 0, \quad \sigma_{22} = 0 \quad \text{at } x_2 = 0, \tag{7.9}$$

allow one to obtain the corresponding dispersion equation (but here there is no real dispersion). The Rayleigh wave is the real solution of this equation of dispersion. This mode propagates in the x_1 direction with a speed noted c_R and is evanescent in the x_2 vertical direction. So, it is a surface wave.

This speed c_R is given by the real solution of the so-called *Rayleigh equation* (Achenbach (1973); Dieulesaint and Royer (2000)):

$$D(c_R; c_T, c_L) \equiv 4 \sqrt{1 - \frac{c_R^2}{c_T^2}} \sqrt{1 - \frac{c_R^2}{c_L^2}} - \left(2 - \frac{c_R^2}{c_T^2}\right)^2 = 0. \qquad (7.10)$$

The displacement of the matter particles during the Rayleigh wave propagation is elliptic with a magnitude decreasing in depth with the variable x_2. The two components (u_1, u_2) of the displacement vector in the so-called sagittal plane are given by $[x = x_1]$

$$u_1 = U(x_2) \sin(k_R\, x - \omega\, t), \quad u_2 = V(x_2) \cos(k_R\, x - \omega\, t), \qquad (7.11)$$

with

$$\begin{cases} U(x_2) = -k_R\, A\left(\exp(-\alpha_L\, x_2) - \dfrac{2\, \alpha_L\, \alpha_T}{k_R^2 + \alpha_T^2} \exp(-\alpha_T\, x_2) \right), \\[2mm] V(x_2) = -\alpha_L\, A\left(\exp(-\alpha_L\, x_2) - \dfrac{2\, k_R^2}{k_R^2 + \alpha_T^2} \exp(-\alpha_T\, x_2) \right), \end{cases} \qquad (7.12)$$

where A is a real constant which has the dimension of a squared length (relatively small compared to the wavelength in linear elasticity), and

$$k_R = \frac{\omega}{c_R}, \qquad k_L = \frac{\omega}{c_L}, \qquad k_T = \frac{\omega}{c_T}, \qquad (7.13)$$

are respectively the wave numbers of Rayleigh wave, of longitudinal waves and of transverse waves. In addition,

$$\alpha_L = \sqrt{k_R^2 - k_L^2}, \qquad \alpha_T = \sqrt{k_R^2 - k_T^2}, \qquad (7.14)$$

are representative of the evanescence of longitudinal and transverse components which compose the Rayleigh wave. Let us remark that the evanescence of longitudinal and transverse waves is a necessary condition for the existence of the Rayleigh wave. With this condition we have the inequalities

$$k_L < k_T < k_R \quad \text{and} \quad c_L > c_T > c_R. \qquad (7.15)$$

7.2.2 *Conservation of wave momentum for Rayleigh surface waves*

The local *conservation law of wave momentum* is given by (*cf.* Chapter 3)

$$\frac{\partial}{\partial t} \mathbf{p}^w - div\, \mathbf{b}^w = \mathbf{0}, \qquad (7.16)$$

with (T denotes transposition)

$$\mathbf{p}^w = -\rho_0 \frac{\partial \mathbf{u}}{\partial t} \cdot (\nabla \mathbf{u})^T, \qquad \mathbf{b}^w = -(K - W)\, \mathbf{1} - \boldsymbol{\sigma} \cdot (\nabla \mathbf{u})^T, \qquad (7.17)$$

respectively the *wave momentum* and the so-called *Eshelby stress*. In Cartesian tensor notation, these read

$$p_i^w = -\rho_0 \, \dot{u}_k \, u_{k,i}, \qquad b_{ji}^w = -(L \, \delta_{ji} + \sigma_{jk} \, u_{k,i}). \tag{7.18}$$

Equation (7.16) is accompanied by the *conservation of energy* which reads here in local form as

$$\frac{\partial}{\partial t}(K + W) - \frac{\partial}{\partial x_j}(\sigma_{ji} \, \dot{u}_i) = 0, \tag{7.19}$$

where the Hamiltonian density $H = K + W$ is such that $H = \mathbf{p}^w \cdot \dot{\mathbf{u}} - L$.

Had we not known Noether's theorem, we could have simply taken the product of (7.2) to the right by $(\nabla \mathbf{u})^T$ and obtained (7.16) after some manipulations assuming the dependence of the energy W on strains. Similarly, (7.19) is obtained by application of Noether's theorem for time translations or by direct computation following the scalar multiplication of (7.2) by the velocity $\dot{\mathbf{u}}$.

To proceed further, we now consider a part of the previous semi-infinite elastic medium of unit thickness in the transverse direction and defined in the sagittal plane by the domain $D = [x_0, x_0 + \lambda_R] \times [0, +\infty)$ — see coloured Plate 7.2 — which is a semi-infinite vertical band of longitudinal width equal to one (Rayleigh) wavelength.

The conservation law (7.16) of wave momentum integrated over the material domain D gives by use of the divergence theorem

$$\frac{d}{dt} \int_D -\rho_0 \frac{\partial \mathbf{u}}{\partial t} \cdot (\nabla \mathbf{u})^T \, dV$$
$$- \int_{\partial D} \mathbf{n} \cdot \left(-(K - W)\mathbf{1} - \boldsymbol{\sigma} \cdot (\nabla \mathbf{u})^T \right) dS = 0, \tag{7.20}$$

where dV is an element of volume, dS an element of surface, ∂D is the boundary of D and $\mathbf{n}$ is the external unit vector to ∂D.

As the surface integral is simplified by the free boundary conditions, the vanishing condition at infinity, and the x_1-periodicity of the solution, equation (7.20) reduces to (here $\mathbf{n} = -\mathbf{e}_2$ at $x_2 = 0$)

$$\frac{d}{dt} \int_D -\rho_0 \frac{\partial \mathbf{u}}{\partial t} \cdot (\nabla \mathbf{u})^T \, dV = \int_{x_0}^{x_0 + \lambda_R} (K - W)(x, x_2 = 0) \, dx \, \mathbf{e}_2. \tag{7.21}$$

7.2.3 Quasi-particles associated with Rayleigh surface waves

The evaluation of the volume integral in the left-hand side of (7.21) (in fact a surface integral with unit thickness in the direction orthogonal to the sagittal plane) is easy but lengthy. It goes as follows. This quantity reads

$$\int_D -\rho_0 \frac{\partial \mathbf{u}}{\partial t} \cdot (\nabla \mathbf{u})^T \, dv$$

$$= \int_0^\infty \int_{x_0}^{x_0 + \lambda_R} -\rho_0 \left(u_{1,1} \, u_{1,t} + u_{2,1} \, u_{2,t} \right) dx_1 \, dx_2 \, \mathbf{e}_1 \qquad (7.22)$$

$$+ \int_0^\infty \int_{x_0}^{x_0 + \lambda_R} -\rho_0 \left(u_{1,2z} \, u_{1,t} + u_{2,2} \, u_{2,t} \right) dx_1 \, dx_2 \, \mathbf{e}_2.$$

The first term in the right-hand side of this vectorial expression contains terms in $\cos^2(k_R x - \omega t)$ and $\sin^2(k_R x - \omega t)$, so that its contribution is *not* zero by integration along one wavelength. The evaluation of this term is lengthy but it allows one to obtain the expression of the mass M_R (see below). The second term in the right-hand side contains only terms in $\cos(k_R x - \omega t) \sin(k_R x - \omega t)$; so its contribution is zero by integration along one wavelength. As a result equation (7.22) has only a nonzero component along $\mathbf{e}_1$. Furthermore, it remains to evaluate the right-hand side of equation (7.21). This contribution has no contribution along $\mathbf{e}_1$. There remains to evaluate the $\mathbf{e}_2$-component, i.e.

$$\int_{x_0}^{x_0 + \lambda_R} (K - W)(x_1, x_2 = 0) \, dx \, \mathbf{e}_2$$

$$= \frac{1}{2} \rho_0 \int_{x_0}^{x_0 + \lambda_R} \left\{ (u_{1,t}^2 + u_{2,t}^2) - \left[c_L^2 \, (u_{1,1}^2 + u_{2,2}^2) \right. \right.$$

$$+ 2 \, (c_L^2 - 2 \, c_T^2) \, u_{1,1} \, u_{2,2} + c_T^2 \, (u_{1,2}^2 + u_{2,1}^2) \qquad (7.23)$$

$$\left. \left. + 2 \, c_T^2 \, u_{1,2} \, u_{2,1} \right] \right\} (x_2 = 0) \, dx_1 \, \mathbf{e}_2$$

It is easy to show that this whole term vanishes as a consequence of the "dispersion" relation (7.10). This coincides with a result of the kinematic theory of waves (Lighthill–Whitham) where it is shown that the averaged value of the Lagrangian $L = K - W$ over a period of motion vanishes when the *linear* dispersion relation is satisfied as the above expression is essentially $D(c_R; c_T, c_L) \, \overline{A}^2$, in a classical formalism $(\overline{A} = A \, k_R)$; in this regard compare to Whitham (1974), and more recently Maugin (2007). We have thus checked that the two sides of equation (7.21) projected along $\mathbf{e}_2$ vanish identically in the considered semi-infinite band. As to the x_1-component, it is obtained as (*cf.* Rousseau and Maugin (2011a))

$$\int_D -\rho_0 \frac{\partial \mathbf{u}}{\partial t} \cdot (\nabla \mathbf{u})^T \, dV = M_R \, c_R \, \mathbf{e}_1. \qquad (7.24)$$

In this equation, c_R which is the Rayleigh wave speed, is now viewed as a speed of "information". Then, M_R (which is homogeneous to a mass per unit of length in the direction $\mathbf{e}_3$) is viewed as a "mass", not of matter but of "information". This one is representative of an "average acoustic energy" or of a certain amount of "acoustic information", and so defines the quasi-particle of Rayleigh. It is found to be given by

$$M_R = \rho_0 \, \pi \, f(k_R, \alpha_L, \alpha_T) \, A^2, \qquad (7.25)$$

with (here $k_R \, A$ will be the amplitude of the displacement while A was that of the potential):

$$f(k_R; \alpha_L, \alpha_T) = k_R^2 \left[\frac{k_R^2 + \alpha_L^2}{4 \, k_R \, \alpha_L} + \frac{2 \, \alpha_L^2 \, \alpha_T \, k_R^2}{k_R \, (k_R^2 + \alpha_T^2)^2} \right.$$
$$\left. + \frac{2 \, \alpha_L \, \alpha_T \, k_R^2 + \alpha_L^2 \, (k_R^2 + \alpha_T^2)}{k_R \, (k_R^2 + \alpha_T^2) \, (\alpha_L + \alpha_T)} \right]. \qquad (7.26)$$

This parameter f depends on the mechanical characteristics of the elastic medium and bears the print of the boundary conditions. It is the quantity that truly characterizes the quasi-particle of Rayleigh.

In summary, we have found that the motion of the relevant quasi-particle is governed by the equation

$$\frac{d}{dt}(M_R \, c_R) = 0, \qquad (7.27)$$

that represents the *inertial* motion (conservation of momentum in Newton's sense) of the quasi-particle of "mass" M_R along the surface of the limiting plane. So, we have obtained a Newtonian formalism of the Rayleigh wave propagation viewed as a "particle". It remains to check *independently* that the equation of energy reduces to conservation in time of a kinetic energy. Indeed, the canonical conservation laws form a true *thermo*-mechanics, in the sense that the energy conservation should always be considered in parallel with that of canonical momentum. We did not do it here. However, with integration over the semi-infinite band $D = [x_0, x_0 + \lambda_R] \times [0, +\infty)$ of equation (7.19), we would show that the corresponding total energy E takes a Newtonian form such as

$$E = \frac{1}{2} \, M_R \, c_R^2. \qquad (7.28)$$

This is obviously conserved during the motion. That is, the energy of the "point-like" Rayleigh quasi-particle appears as purely kinetic in agreement with equation (7.27) while in the continuum description the energy is made of kinetic *and* potential (elastic) parts; but M_R still contains all the relevant information. The situation is even more spectacular in the electroelastic case treated in Chapter 8.

7.2.4   *The influence of surface energy*

The perturbation obtained by considering a *surface energy* F along the top surface $x_2 = 0$ is of special interest for technological applications at a *nanoscale*. This case was considered in Rousseau and Maugin (2011a). The relevant results are as follows. The free boundary vectorial condition (7.3) at $x_2 = 0$, is now replaced by the two scalar conditions:

$$\sigma_{21} = 0, \qquad \sigma_{22} = -2\,F\,\frac{\partial^2 u_2}{\partial x_1^2}. \qquad (7.29)$$

The situation and solution are summarized in the coloured Plate 7.3.

A small parameter $\varepsilon_F = F\,k_T/\mu$ or $\varepsilon_F(\omega) = (F/\mu\,c_T)\,\omega$ which compares surface and bulk elasticity coefficients is introduced. The perturbed dispersion relation at order ε now becomes

$$D(c_R; c_T, c_L) + \varepsilon_F\,\frac{c_{RF}}{c_T}\,\sqrt{1 - \frac{c_{RF}{}^2}{c_L{}^2}} = 0, \qquad (7.30)$$

of which the solution is the corrected wave speed c_{RF} of the Rayleigh mode as obtained by Vlasie–Belloncle and Rousseau (2006):

$$c_{RF}(\omega) = c_R\left(1 - \varepsilon_F(\omega)\,\alpha_F\right), \qquad (7.31)$$

where α_F is a complicated negative constant (in terms of k_R, k_T and k_L) of order unity whose precise expression is irrelevant. It was observed that with increasing frequency from zero to a critical value corresponding to the limit of existence of the Rayleigh wave, c_{RF} increases from c_R (no surface energy) to c_T. Since there is no dissipation and no driving force acting on the quasi-particle it is found that for the volume integral of (7.16) we obtain a Newtonian-like equation of motion in the form

$$\frac{d}{dt}(M_{RF}\,c_{RF}) = 0, \qquad (7.32)$$

where the "mass" M_{RF} is given by

$$M_{RF} = M_R(1 + \varepsilon_F\,\alpha_F). \qquad (7.33)$$

Since $\alpha_F < 0$, this "mass" is smaller than the one associated with the pure Rayleigh SAW, while $c_{RF} \geq c_R$.

7.2.5   *The case of leaky surface waves*

We evoke this pathological case for the sake of completeness. Consider that at $x_2 = 0$ the elastic medium is in contact with a low-density inviscid fluid medium which occupies the half-space $x_2 < 0$. The Rayleigh wave associated with the semi-infinite elastic medium is then perturbed by the existence of the fluid and it becomes a *leaky* Rayleigh wave. The boundary conditions at $x_2 = 0$ are now written as

$$\sigma_{21} = 0, \qquad \sigma_{22} = -p, \qquad [u_1] = 0, \qquad (7.34)$$

where p is the acoustic pressure in the fluid and $[\ldots]$ denotes the jump through the fluid/solid interface. Only displacement u_1 is required to satisfy a boundary condition, since u_2 is arbitrary as the fluid considered is *not* viscous. The situation and solution are summarized in the coloured Plate 7.4. The conditions (7.34) lead to the dispersion relation (for this see Gubanov (1945); Brekhovskikh (1960, p. 42))

$$D(c_{lR}; c_T, c_L) - i\,\varepsilon_l \frac{c_{lR}^4}{c_T^4} \frac{\sqrt{1 - c_{lR}^2/c_L^2}}{\sqrt{c_{lR}^2/c_1^2 - 1}} = 0, \qquad (7.35)$$

where $\varepsilon_l = \rho_1/\rho_0$ (ratio between the mass densities of the fluid and of the solid) will be supposed small (this is the case for, e.g., a water/metal interface), and c_1 is the speed of sound in the fluid medium. Thus the second term in equation (7.35) appears as a perturbation of the Rayleigh equation (7.10) with speed c_{lR}; note the presence of the imaginary number. The leaky Rayleigh wave is then the *complex* solution of the dispersion equation (7.35). This solution can be written in perturbed form as

$$c_{lR} = c_R \left(1 + i\,\varepsilon_l\,\alpha_l\right), \qquad (7.36)$$

wherein (this can be obtained by differentiating (7.35) and identifying α_l)

$$\alpha_l = \frac{(c_R^2/c_T^2)\sqrt{1 - c_R^2/c_L^2}}{4\sqrt{\dfrac{c_R^2}{c_1^2} - 1}\left[1 - \dfrac{\sqrt{1 - c_R^2/c_L^2}}{\sqrt{1 - c_R^2/c_T^2}} - \dfrac{c_T^2}{c_L^2}\dfrac{\sqrt{1 - c_R^2/c_T^2}}{\sqrt{1 - c_R^2/c_L^2}}\right]}. \qquad (7.37)$$

The latter depends on the speed (c_R) of the unperturbed Rayleigh wave, the transverse (c_T) wave and the longitudinal (c_L) wave in the solid and of the sound speed c_1 in the fluid. This parameter is a *positive* constant. So, the imaginary part of c_{lR} indicates that, in the solid, the wave propagates obliquely from the solid to the fluid and vanishes at infinity. At the same time, in the fluid the wave propagates obliquely and increases at infinity. That clearly is physically untenable. This defines the "pathological" notion of *leaky Rayleigh wave*. Still we consider the possible integration of the corresponding equation (7.16) over the volume

$\Omega = [x_{10}, x_{10} + \lambda^w] \times [0, +\infty) \times [0, 1]$. On computing the total wave momentum in the x_1 direction we find the Newtonian-like equation of inertial motion

$$\frac{d}{dt}(M_{lR}\, c_{lR}) = 0, \tag{7.38}$$

where c_{lR} is the speed of the leaky Rayleigh wave and M_{lR} is its "mass". This conservation of momentum is verified for any ε_l and especially when $\varepsilon_l \to 0$ (i.e. the mass density of the fluid goes to zero) and consequently, by continuity, equation (7.38) can be written as

$$M_{lR}\, c_{lR} = M_R\, c_R, \tag{7.39}$$

so that, to first order, the "mass" of the quasi-particle of the leaky Rayleigh wave is given by the *complex* expression

$$M_{lR} = M_R\,(1 - i\,\varepsilon_l\,\alpha_l). \tag{7.40}$$

The imaginary part of this "mass" is negative and describes the re-emitting of a part of the "information" from the solid to the fluid. This is the "leak" referred to in the name of the wave. In the fluid we obtain an accumulation of "information" at infinity, which again is untenable. The reached nonsensical situation can only be solved by the theory of *bounded beams* in the framework of wave theory (Bertoni and Tamir (1973)). The question of how to remedy this deficiency in the quasi-particle framework is unsettled.

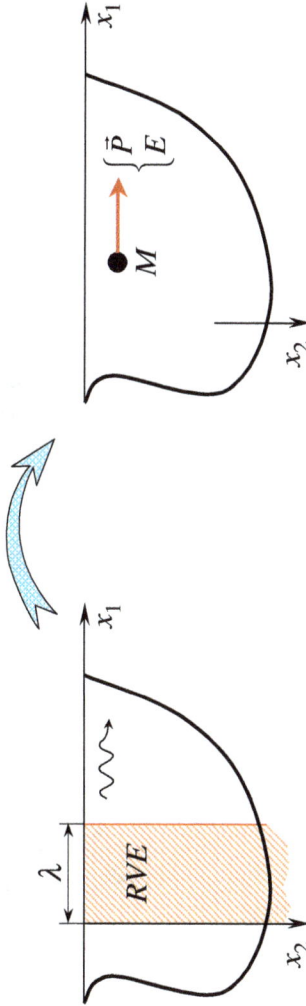

Plate 7.1   Passing from SAWs to quasi-particles.

Standard Rayleigh SAW

Dispersion relation:

$$D(c_R; c_T, c_L) \equiv 4\sqrt{1-(c_R/c_T)^2}\sqrt{1-(c_R/c_L)^2} - \left(2-(c_R/c_T)^2\right)^2 = 0$$

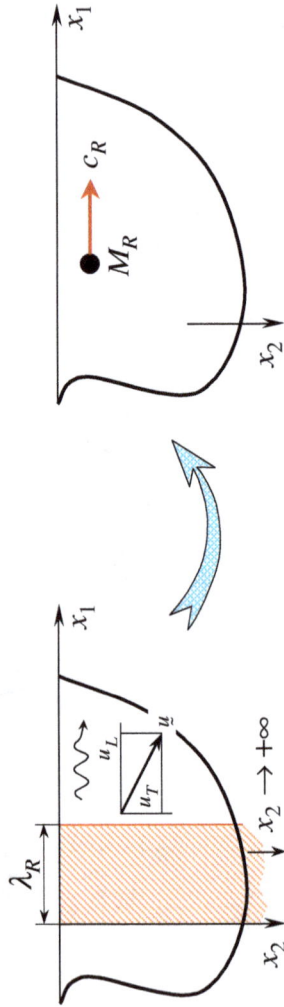

$$\frac{\mathrm{d}}{\mathrm{d}t}(M_R\,c_R)=0, \quad M_R = \rho_0\,\pi\,f(k_R,\alpha_L,\alpha_T)U^2, \quad E_R = \frac{1}{2}M_R\,c_R^2$$

Plate 7.2   The case of standard Rayleigh SAW.

Rayleigh SAW + surface energy

Dispersion relation:

$$D(c_{RF}; c_T, c_L) + \varepsilon_F(c_{RF}/c_F)\sqrt{1-(c_{RF}/c_L)^2} = 0$$

surface energy $F \rightarrow \varepsilon_F$

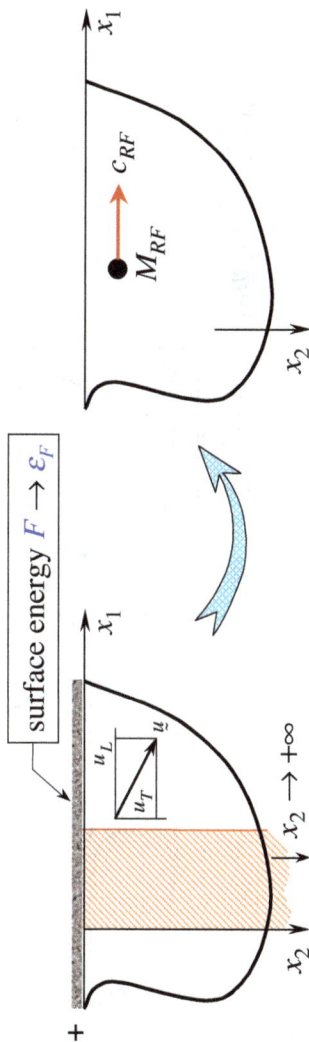

$$\frac{\mathrm{d}}{\mathrm{d}t}(M_{RF}\,c_{RF}) = 0, \quad M_{RF} = M_R(1+\varepsilon_F\,\alpha_F), \quad c_{RF} = c_R(1-\varepsilon_F(\omega)\alpha_F)$$

$$(\alpha_F < 0)$$

Plate 7.3 The case of Rayleigh SAW with surface energy.

Leaky Rayleigh wave

Dispersion relation:

$$D(c_{\ell R}; c_T, c_L) - i\varepsilon_\ell \,(c_{\ell R}/c_T)^4 \sqrt{1-(c_{\ell R}/c_L)^2}\,\sqrt{(c_{\ell R}/c_T)^2 - 1} = 0$$

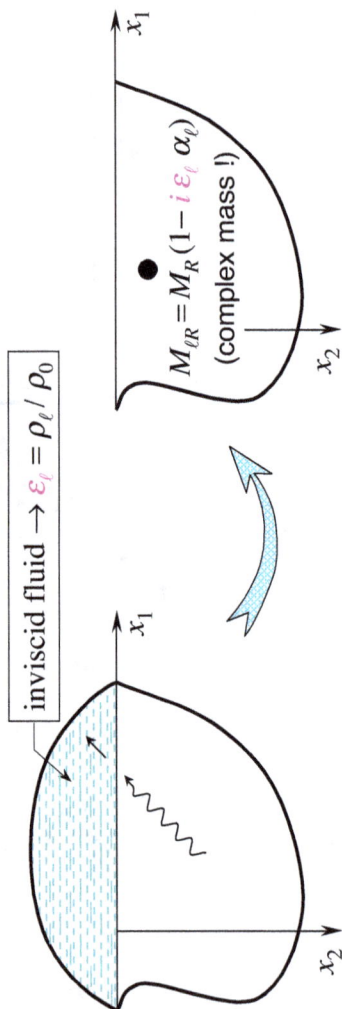

inviscid fluid $\rightarrow \varepsilon_\ell = \rho_\ell / \rho_0$

$$M_{\ell R} = M_R(1 - i\,\varepsilon_\ell\,\alpha_\ell)$$
(complex mass !)

Pathological situation $\rightarrow$ Solution: bounded beams

Plate 7.4   The case of leaky Rayleigh waves.

7.3   The case of Love waves

7.3.1   *The Love SAW solution*

This solution was given by A.E.H. Love (1863–1940) in Chapter 11, "Theory of propagation of seismic waves", of a prize-winning essay published in book form (Love (1911)). In his thoughtful review Wilson (1914) speaks of this essay as a "massive intricate analysis interspersed with readable contents". Fortunately, this new surface solution is relatively simple. We recall about it only the main elements that can be found in greater detail in general books on elastic waves (e.g., Achenbach (1973)).

With the same notation as in previous sections, the considered composite structure is made of the semi-infinite substrate (body numbered 2) occupying the region $x_2 > 0$, and of a superimposed layer of finite thickness d (body numbered 1) occupying the region $-d < x_2 < 0$. Both bodies are linear isotropic elastic bodies with constant mass densities and shear elasticity coefficients accordingly noted ρ_1 and ρ_2 and μ_1 and μ_2. The associated transverse elastic wave speeds are defined by

$$c_{T_1} = \sqrt{\frac{\mu_1}{\rho_1}}, \qquad c_{T_2} = \sqrt{\frac{\mu_2}{\rho_2}}. \tag{7.41}$$

The Love SAW is a dynamical (out of (sagittal) plane) SH motion with displacement such as

$$\mathbf{u} = u_3(x_1, x_2, t)\, \mathbf{e}_3, \tag{7.42}$$

with propagation along the x_1 direction and amplitude monotonically decreasing in depth along x_2. The superimposed layer plays the role of a wave guide. In both regions the standard equation of motion (with the appropriate labelling for each region) reads

$$\frac{\partial}{\partial t}(\rho\, \dot{\mathbf{u}}) - div\, \boldsymbol{\sigma} = \mathbf{0}, \tag{7.43}$$

in the absence of body force, where $\dot{\mathbf{u}} = \partial \mathbf{u}/\partial t$ and $\boldsymbol{\sigma}$ denotes the symmetric Cauchy stress given in Cartesian tensor components by equation (7.1) where λ and μ are Lamé coefficients of isotropic elasticity.

Because of the chosen specific displacement field, Equation (7.43) is reduced to the form (with the appropriate labelling for each region)

$$\frac{\partial^2 u_3}{\partial t^2} = c_T^2 \left(\frac{\partial^2 u_3}{\partial x_1^2} + \frac{\partial^2 u_3}{\partial x_2^2} \right) = c_T^2\, \Delta u_3, \tag{7.44}$$

where Δ is the two-dimensional Laplacian operator in the sagittal plane (x_1, x_2).

Boundary and junction conditions

The top surface of the superimposed layer is free of load, hence with vanishing traction with unit outward oriented normal $\mathbf{n}$ (*cf.* Left part in coloured Plate 7.5):

$$\mathbf{n} \cdot \boldsymbol{\sigma} = \mathbf{0} \quad \text{at } x_2 = -d, \tag{7.45}$$

while perfect contact is assumed between the two bodies at the interface, thus with continuous displacement and traction:

$$[\mathbf{u}] = \mathbf{0}, \qquad [\mathbf{n} \cdot \boldsymbol{\sigma}] = \mathbf{0} \quad \text{at } x_2 = 0, \tag{7.46}$$

where the symbolism $[\ldots]$ denotes the jump of its enclosure. For our particular dynamical problem (7.42), Equations (7.45) and (7.46) take the form

$$\sigma_{2i}(1) = 0, \quad \text{or } u_{3,2}(1) = 0 \quad \text{at } x_2 = -d, \tag{7.47}$$

and

$$u_3(1) = u_3(2), \qquad \mu_1 \, u_{3,2}(1) - \mu_2 \, u_{3,2}(2) = 0 \quad \text{at } x_2 = 0. \tag{7.48}$$

Because of the guiding nature of the superimposed layer the solution of Equations (7.44) for bodies 1 and 2 is looked for in the *a priori* form $(x = x_1)$

$$u_3 = A \, \exp(-k_2 \, x_2) \, \exp\left(i \, (k \, x - \omega \, t)\right) \quad \text{for } x_2 > 0, \tag{7.49}$$

$$u_3 = \left(B_1 \, \sin(k_B \, x_2) + B_2 \, \cos(k_B \, x_2)\right) \exp\left(i \, (k \, x - \omega \, t)\right)$$
$$\text{for } -d < x_2 < 0, \tag{7.50}$$

with

$$k_B := k \, \sqrt{\frac{c^2}{c_{T_1}^2} - 1}, \qquad c = \frac{\omega}{k}. \tag{7.51}$$

The conditions $(7.48)_1$, $(7.48)_2$ and (7.47) yield the relations

$$B_2 = A, \tag{7.52}$$

$$\mu_1 \, B_1 \, k_B + \mu_2 \, k_2 \, A = 0, \tag{7.53}$$

$$B_1 \, \cos(k_B \, d) + B_2 \, \sin(k_B \, d) = 0. \tag{7.54}$$

The solution of this system yields

$$B_1 = - \tan(k_B \, d) \, B_2 = - \tan(k_B \, d) \, A, \tag{7.55}$$

and

$$\tan\left(\sqrt{c^2/c_{T_1}^2 - 1} \, k \, d\right) - \frac{\mu_2}{\mu_1} \frac{\sqrt{1 - c^2/c_{T_2}^2}}{\sqrt{c^2/c_{T_1}^2 - 1}} := D_L(c, k) = 0, \tag{7.56}$$

$$k_2 = k \, \sqrt{1 - c^2/c_{T_2}^2}. \tag{7.57}$$

Equation (7.56), $D_L(c, k) = 0$, is the *dispersion relation* of Love SAWs. It requires that $c > c_{T_1}$ and $c < c_{T_2}$, so that

$$c_{T_1} < c < c_{T_2}. \tag{7.58}$$

It can be said that that the Love SAW is supersonic with respect to the shear-wave speed of the superimposed layer and subsonic with respect to that of the substrate. The layer must be "*slow*" compared to the substrate.

We can also introduce a more vivid notation akin to a relativistic one:

$$\beta_1 = \frac{c}{c_{T_1}} > 1, \qquad \beta_2 = \frac{c}{c_{T_2}} < 1, \tag{7.59}$$

and (kind of Lorentz factors)

$$\gamma_1 = \frac{1}{\sqrt{\beta_1^2 - 1}} < 1, \qquad \gamma_2 = \frac{1}{\sqrt{1 - \beta_2^2}} > 1. \tag{7.60}$$

But an essential difference with a typical relativistic (Lorentz–Einstein) motion here is that we witness the existence of *both* upper and lower bounds for the velocity, hence the first of (7.60) which can be viewed as defining an "anti-Lorentz" factor. In partial conclusion, we note that, contrary to Rayleigh SAWs, Love SAWS are *dispersive* and *multimode* scalar waves with the mentioned peculiarity for their possible propagation velocity.

7.3.2 Conservation of wave momentum and energy

Following the application of Noether's theorem, one can associate with Equation (7.43) the following local equations of conservation of wave-momentum and energy at each point of the two elastic bodies of the Love-wave setting:

$$\frac{\partial}{\partial t} \mathbf{p}^w - div\, \mathbf{b} = \mathbf{0}, \tag{7.61}$$

and

$$\frac{\partial}{\partial t} E - \nabla \cdot \mathbf{Q} = 0. \tag{7.62}$$

Here

$$\mathbf{p}^w = \left\{ p_i^w = -\rho\, \dot{u}_j\, u_{j,i} \right\},$$
$$\mathbf{b} = -\left(L\, \mathbf{1} + \boldsymbol{\sigma} \cdot (\nabla \mathbf{u})^T \right) = \left\{ b_{ji} = -(L\, \delta_{ji} + \sigma_{jk}\, u_{k,i}) \right\}, \tag{7.63}$$
$$L = K - W, \qquad E = K + W, \qquad \mathbf{Q} = \boldsymbol{\sigma} \cdot \dot{\mathbf{u}}. \tag{7.64}$$

In the present problem with displacement function (7.42), the following special expressions apply:

$$p_1^w = -\rho\, \dot{u}_3\, u_{3,1}, \qquad p_2^w = -\rho\, \dot{u}_3\, u_{3,2}, \qquad K = \frac{1}{2}\rho\, \dot{u}_3^2, \tag{7.65}$$

$$W = \frac{1}{2}\mu\left(u_{3,1}^2 + u_{3,2}^2 \right), \qquad Q_1 = \sigma_{13}\, \dot{u}_3, \qquad Q_2 = \sigma_{23}\, \dot{u}_3. \tag{7.66}$$

7.3.3   *Mass and energy of the associated quasi-particle*

Following previously treated cases, the applied strategy consists in integrating the relevant components of (7.61) — as also Equation (7.62) itself — over an element of volume representative (i.e., RVE) of the studied wave process while accounting for the already deduced wave-like solution with a real propagation factor $\cos(k\,x - \omega\,t)$ replacing the imaginary exponentials. Here this RVE is composed of one wave length λ in the propagation direction x_1, a vertical strip of this width extending from the top surface of the layer at $x_2 = -d$ to infinity in the depth of the substrate, and a unit length in the orthogonal direction x_3, hence the RVE given by (see coloured Plate 7.5)

$$\Omega = [x_0, x_0 + \lambda] \times [-d, +\infty) \times [0, 1]. \tag{7.67}$$

Integration along x_1 and then dividing by λ is nothing but an average over a wavelength.

The global conservation equations to be considered are, from (7.61) and (7.62):

$$\frac{\partial}{\partial t} \int_\Omega p_1^w \, dV - \int_{\partial\Omega} (\mathbf{n} \cdot \mathbf{b} \cdot \mathbf{e}_1) \, dS = 0, \tag{7.68}$$

$$\frac{\partial}{\partial t} \int_\Omega p_2^w \, dV - \int_{\partial\Omega} (\mathbf{n} \cdot \mathbf{b} \cdot \mathbf{e}_2) \, dS = 0, \tag{7.69}$$

and

$$\frac{\partial}{\partial t} \int_\Omega E \, dV - \int_{\partial\Omega} \mathbf{n} \cdot \mathbf{Q} \, dS = 0. \tag{7.70}$$

All terms should be computed on account of the solution (7.49)–(7.50), the already deduced dispersion relation, the condition of periodicity along the propagation direction, and the asymptotic condition in the substrate. These are algebraic calculations involving trigonometric functions that just require some care (details, where needed, are given in the Appendices to Rousseau and Maugin (2014)). Thus,

$$\int_\Omega p_1^w \, dV = \int_0^\lambda \left(\int_{-d}^0 p_1^w \, dx_2 + \int_0^{+\infty} p_1^w \, dx_2 \right) dx_1, \tag{7.71}$$

$$\int_\Omega p_2^w \, dV = \int_0^\lambda \left(\int_{-d}^0 p_2^w \, dx_2 + \int_0^{+\infty} p_2^w \, dx_2 \right) dx_1, \tag{7.72}$$

$$\int_{\partial\Omega} (\mathbf{n}\cdot\mathbf{b}\cdot\mathbf{e}_1)\,dS = \int_0^\lambda (-b_{21}(1))\big|_{x_2=-d}\,dx_1$$
$$+ \int_0^\lambda (b_{21}(1)-b_{21}(2))\big|_{x_2=0}\,dx_1 + OT, \tag{7.73}$$

$$\int_{\partial\Omega} (\mathbf{n}\cdot\mathbf{b}\cdot\mathbf{e}_2)\,dS = \int_0^\lambda (-b_{22}(1))\big|_{x_2=-d}\,dx_1$$
$$+ \int_0^\lambda (b_{22}(1)-b_{22}(2))\big|_{x_2=0}\,dx_1 + OT, \tag{7.74}$$

$$\int_\Omega E\,dV = \int_0^\lambda \left(\int_{-d}^0 E\,dx_2 + \int_0^{+\infty} E\,dx_2\right)\,dx_1, \tag{7.75}$$

$$\int_{\partial\Omega} \mathbf{n}\cdot\mathbf{Q}\,dS = \int_0^\lambda (-Q_2(1))\big|_{x_2=-d}\,dx_1$$
$$+ \int_0^\lambda (Q_2(1)-Q_2(2))\big|_{x_2=0}\,dx_1 + OT, \tag{7.76}$$

where the symbol OT stands for terms that will not contribute due to the periodicity along x_1 (at the left and right sides of the vertical strip) and to vanishing terms at $x_2 \to +\infty$.

It can be shown that

$$\int_{\partial\Omega} (\mathbf{n}\cdot\mathbf{b}\cdot\mathbf{e}_1)\,dS \equiv 0, \qquad \int_{\partial\Omega} (\mathbf{n}\cdot\mathbf{b}\cdot\mathbf{e}_2)\,dS \equiv 0, \tag{7.77}$$

$$\int_\Omega \mathbf{n}\cdot\mathbf{Q}\,dS \equiv 0, \tag{7.78}$$

and

$$P_2 := \int_\Omega p_2^w\,dV \equiv 0, \tag{7.79}$$

so that (7.69) yields $0 \equiv 0$, and we are left with the unique scalar equation of motion in the propagation direction in the inertial form

$$\frac{d}{dt}P_1 = 0, \tag{7.80}$$

where the linear momentum P_1 is obtained as (See equations (A.17)–(A.19) in Rousseau and Maugin (2014))

$$P_1 = P = M_L\,c, \tag{7.81}$$

with a mass

$$M_L = \frac{\rho_1\,\pi\,A^2\,k}{2}\,d\left(1 + \tan^2(k_B\,d) + \frac{\tan(k_B\,d)}{k_B\,d}\right) + \frac{\rho_2\,\pi\,A^2\,k}{2\,k_2}, \tag{7.82}$$

or

$$M_L = \rho_1 \pi^2 A^2 \frac{d}{\lambda} \left(1 + \tan^2(k_B d) + \frac{\tan(k_B d)}{k_B d} \right) + \frac{\rho_2 \pi^2 A^2}{k_2 d} \frac{d}{\lambda}. \qquad (7.83)$$

Equation (7.81) can be viewed as the definition of the linear momentum of a *non-Newtonian quasi-particle* propagating at speed c but with a velocity-dependent mass (proportional to the squared amplitude of the wave form).

The above-obtained results forcefully rely on the following expressions that we give to help the reader who would like to redo all painstaking computations. In particular, the following should be noted for the evaluation of various terms in Equations (7.71) through (7.76) and the proof of Equations (7.77) through (7.79):

$$(-b_{21}(1))\big|_{x_2=-d} = (\sigma_{23}(1)\, u_{3,1}(1))\big|_{x_2=-d} = 0, \qquad (7.84)$$

$$(\sigma_{23}(2)\, u_{3,1}(2) - \sigma_{23}(1)\, u_{3,1}(1))\big|_{x_2=0}$$
$$= -\left(\mu_1 k_B B_1 B_2 + \mu_2 k_2 A^2 \right)\big|_{x_2=0}, \qquad (7.85)$$

(the latter up to the propagation factor), and

$$b_{22}(1)\big|_{x_2=-d} = -(L(1) - \sigma_{23}(1)\, u_{3,2}(1))\big|_{x_2=-d} = -(L(1))\big|_{x_2=-d}, \quad (7.86)$$

$$(b_{22}(1) - b_{22}(2))\big|_{x_2=0} = -\big(L(1) + \sigma_{23}(1)\, u_{3,2}(1)$$
$$- L(2) - \sigma_{23}(2)\, u_{3,2}(2)\big)\big|_{x_2=0}, \qquad (7.87)$$

with $(\alpha = 1, 2)$

$$L(\alpha) + \sigma_{23}(\alpha)\, u_{3,2}(\alpha) = \frac{1}{2} \rho_\alpha \dot{u}_3^2(\alpha) + \frac{1}{2} \mu_\alpha \left(u_{3,2}^2(\alpha) - u_{3,1}^2(\alpha) \right). \qquad (7.88)$$

It is further shown that

$$\int_0^{\lambda = 2\pi/k} b_{22}(1)\big|_{x_2=-d}\, dx_1$$
$$= -\frac{1}{2} \left(\rho_1 \omega^2 - \mu_1 k^2 \right) A^2 \frac{\pi}{k} \left(1 + \tan^2(k_B d) \right), \qquad (7.89)$$

$$P_1(1) = \int_0^{\lambda = 2\pi/k} \int_{-d}^{0} (-\rho_1 \dot{u}_3\, u_{3,1})\, dx_2\, dx_1$$
$$= \frac{\rho_1 \omega \pi A^2}{2 \cos^2(k_B d)} \left(d + \frac{\sin(2 k_B d)}{2 k_B} \right), \qquad (7.90)$$

and

$$P_1(2) = \int_0^{\lambda = 2\pi/k} \int_0^{+\infty} (-\rho_1 \dot{u}_3\, u_{3,1})\, dx_2\, dx_1 = M_2\, c, \qquad (7.91)$$

with

$$M_2 = \frac{\rho_2 \, \pi \, A^2 \, k}{2 \, k_2} = \frac{\rho_2 \, \pi^2 \, A^2}{k_2 \, d} \, \frac{d}{\lambda}, \tag{7.92}$$

from which there follows that the quasi-particle associated with the Love wave has linear momentum

$$P = (M_1 + M_2) \, c = M_L \, c, \tag{7.93}$$

with M_L given by (7.82).

Comments on the expression (7.82)

It is of utmost interest to analyse the various contributions to this mass. The last contribution, related to the substrate, can be written in various equivalent forms as

$$M_2 = \left(\frac{\rho_2 \, \pi^2 \, A^2}{k_2 \, d} \right) \frac{d}{\lambda} = \rho_2 \, \pi \, A^2 \, \frac{k}{2 \, k_2} = \frac{\rho_2 \, \pi \, A^2}{\sqrt{1 - \beta_2^2}}. \tag{7.94}$$

The first form introduces artificially the thickness of the superimposed layer for the purpose of writing all contributions (7.83) with the same factor. The last form, with the notation introduced in (7.59)–(7.60), reminds us of a Lorentzian (relativistic) form; this is typical of a point mass moving with a speed limited by an upper bound (here c_{T_2}). The intermediate form emphasizes the fact that the wave decreases exponentially in depth with a penetration depth — or "skin effect" — of the order of $1/k_2$, the factor $1/2$ being due to the fact that the wave momentum is quadratic in the fields, and thus proportional to the squared exponential of decay of the solution.

The first contribution within the parentheses in (7.83) comprises three terms. The first of these provides a simple expression

$$M_1 = \rho_1 \, \pi \, k \, A^2 \, \frac{d}{2}. \tag{7.95}$$

This relates to the superimposed layer per se. This compares to the second expression in (7.84). The thickness d indicates an integration over the thickness, with a factor $1/2$ due again to the quadratic expression of the wave momentum. It obviously coincides with the expression computed for an isolated elastic layer carrying an SH wave (*cf.* Appendix B in Rousseau and Maugin (2014)).

The lower bound in speed (provided by c_{T_1}) intervenes in the expression (7.83) *via* the terms in $\tan(k_B \, d)$ which involve the dispersion that is characteristically introduced by the wave coupling between the superimposed layer and the substrate. The whole complexity and originality of the present case stems from these terms. For small $k_B d$'s only the first order contribution will survive while the second-order term will disappear altogether. That is why the limit obtained when the layer thickness d becomes much smaller than the wavelength λ of the wave is enlightening. In this line of thought, the alternate expression (7.83) emphasizes the role played by the ratio d/λ in the exact Love solution.

The energy expression associated with the inertial motion (7.80) was evaluated in Appendix A of Rousseau and Maugin (2014). The remarkable result here is that the total energy is exactly given by the expression

$$E_L = M_L\, c^2, \qquad (7.96)$$

with M_L given by (7.82). Obtaining this result makes use of the dispersion relation (7.56). Expression (7.96) has an "Einsteinian" outlook, but, as noted before, the "mass" M_L is much more complicated than in the Lorentz-Einstein mechanics of point-mass particles because there exist not only a limiting upper speed of propagation (c_{T_2}) but also a limiting lower speed (c_{T_1}) as indicated by the inequalities (7.58).

7.3.4 *Summary of this section*

The case of surface-wave motion treated in the present section provides the missing link between the original work devoted to the case of Rayleigh waves (Rousseau and Maugin (2011a)) and the work that considered the Murdoch type of surface waves (Rousseau and Maugin (2012b)) and reported in the next section. The great originality in the present case is the non-Newtonian dynamics of the obtained quasi-particle. But it is not that of a relativistic (Lorentz–Einstein) point particle because, as compared to the latter, it must account for the existence of *both* upper and lower bounds for the velocity of propagation, from which there follow rather original expressions for the corresponding mass and energy. Both of these duly account for the dispersion relation of the wave-like motion. This can be compared with the apparently simpler Murdoch case of the next section by taking a thin-film limit (See Paragraph 7.4.6).

7.4 The case of Murdoch waves

7.4.1 *Definition of Murdoch waves*

Another purely mechanical model that also involves a unique SH displacement is the one introduced by Murdoch (1976) after his original work on material boundaries (Gurtin and Murdoch (1975)). This model is particularly interesting because (i) of its purely mechanical nature, (ii) of its relative simplicity with a *dispersive monomode* of propagation only [thus much simpler than the *dispersive multimode* Love SAWs studied in Section 7.3 that need the consideration of a superimposed layer of small but *finite* thickness], and (iii) it lends itself to remarkably simple computations as shown in the rest of this Section. We will first provide a reminder on the model (the present paragraph) and the main features of Murdoch linear surface waves in Paragraph 7.4.2. Paragraph 7.4.3 presents the canonical equations of conservation associated with Murdoch surface waves. These equations are next exploited to yield the quasi-particle re-interpretation of Murdoch SAWs.

We remind the reader that the Murdoch SAWs correspond to a propagation of SAWs on a structure made of an elastic half-space covered by a perfectly glued material surface. The latter possesses its own mass, inertia, and elasticity. This *a priori* is more complicated than the model briefly studied in Paragraph 7.2.4. The considered material surface has a vanishing thickness. It can of course be thought of as a layer of finite thickness that has been squeezed to zero thickness while keeping its essential properties. This corresponds to the *very thin plate limit* considered by Tiersten (1969) in his study of signal-processing devices (wavelength of signal much larger than thickness). The relevant field equations and boundary conditions are as follows.

For $x_2 > 0$ (semi-infinite elastic space)

$$\rho_0 \frac{\partial^2 u_i}{\partial t^2} = \frac{\partial}{\partial x_j} \sigma_{ji}, \qquad (7.97)$$

For $x_2 < 0$ we have a vacuum.

At $x_2 = 0$ (material surface with mass, inertia, and elasticity)

$$\hat{\rho}_0 \frac{\partial^2 \hat{u}_i}{\partial t^2} = \frac{\partial}{\partial \hat{x}_j} \hat{\sigma}_{ji} - n_j \sigma_{ji}^+, \qquad (7.98)$$

where superimposed carets refer to quantities related to the surface of unit outward oriented normal n_j. Thus $\hat{\sigma}_{ji}$ is a surface stress while $\sigma_{ji}^+ = \lim \sigma_{ji}$,

$x_2 \to 0^+$, is the three-dimensional stress from the body. Mass density $\hat{\rho}_0$ is per unit surface. The problem situation and solution are summarized in the coloured Plate 7.6.

With a wave problem depending only on the coordinates (x_1, x_2) in the sagittal plane Π_S and a purely SH displacement u_3 in the x_3-direction orthogonal to Π_S, the relevant components of equations (7.41) and (7.42) are given by

$$\rho_0 \frac{\partial^2 u_3}{\partial t^2} = \frac{\partial}{\partial x_1} \sigma_{13} + \frac{\partial}{\partial x_2} \sigma_{23}, \qquad x_2 > 0, \tag{7.99}$$

and

$$\hat{\rho}_0 \frac{\partial^2 \hat{u}_3}{\partial t^2} = \frac{\partial}{\partial \hat{x}_1} \hat{\sigma}_{13} + \sigma_{23}, \qquad x_2 = 0. \tag{7.100}$$

The stress components are given by

$$\sigma_{ji} = \frac{\partial W}{\partial e_{ji}}, \qquad \hat{\sigma}_{ji} = \frac{\partial \widehat{W}}{\partial \hat{e}_{ji}}, \tag{7.101}$$

where, in the linear case, the bulk and surface energies are given by

$$W = \frac{\lambda}{2} I_1^2 + \mu I_2, \qquad I_\alpha = \text{trace } \mathbf{e}^\alpha, \qquad \mathbf{e} = (\nabla \mathbf{u})_S, \tag{7.102}$$

and

$$\widehat{W} = \frac{\hat{\lambda}}{2} \hat{I}_1^2 + \hat{\mu} \hat{I}_2, \qquad \hat{I}_\alpha = \text{trace } \hat{\mathbf{e}}^\alpha. \tag{7.103}$$

All computations done, we obtain thus the set of two equations:

$$\frac{\partial^2 u_3}{\partial t^2} = c_T^2 \frac{\partial}{\partial x_1} u_{3,1} + c_T^2 \frac{\partial}{\partial x_2} u_{3,2} = c_T^2 \, \Delta u_3, \quad x_2 > 0, \tag{7.104}$$

and

$$\frac{\partial^2 \hat{u}_3}{\partial t^2} = \hat{c}_T^2 \frac{\partial^2 \hat{u}_3}{\partial x_1^2} + \hat{c}_T^2 \, k_a u_{3,2}, \quad \text{at } x_2 = 0, \tag{7.105}$$

where we have set

$$c_T^2 = \frac{\mu}{\rho_0}, \qquad \hat{c}_T^2 = \frac{\hat{\mu}}{\hat{\rho}_0}, \qquad k_a = \frac{\mu}{\hat{\mu}}, \tag{7.106}$$

and Δ now denotes the 2D Laplacian operator in Π_S.

Here k_a is a characteristic wave number (inverse of a characteristic length) that is introduced to compare the remaining two shear-elasticity coefficients.

Furthermore, we must check the condition of perfect gluing of the material interface on the body

$$u_3(x_1, x_2 = 0; t) = \hat{u}(x_1; t), \tag{7.107}$$

and the radiation condition far from the surface in the substrate:

$$u_3(x_1, x_2 \to \infty; t) = 0. \tag{7.108}$$

Remark 7.1 (Remark on a variational formulation). We leave as an exercise to the reader the proof that the above given system of field equations can be derived from the following Lagrangian density per unit length in the x_1-direction and unit thickness in the x_3-direction:

$$L = \frac{1}{2} \left(\hat{\rho}_0\, \hat{u}_{3,t}^2 - \hat{\mu}\, \hat{u}_{3,1}^2 \right) + \int_0^\infty \frac{1}{2} \left(\rho_0\, u_{3,t}^2 - \mu\, \Phi \right) dx_2,$$

$$\Phi := u_{3,1}^2 + u_{3,2}^2. \tag{7.109}$$

This expression accounts for the junction condition (7.107) at $x_2 = 0$ (*cf.* Rousseau and Maugin (2012b)).

7.4.2 *Murdoch SAW linear solution*

This is a solution of the form

$$u_3(x_1, x_2; t) = A\, \exp(-\alpha\, x_2)\, \exp\left(i\, (k_1\, x_1 - \omega\, t) \right). \tag{7.110}$$

The resulting dispersion relation will be a relation $D(\omega, k_1) = 0$ or $\overline{D}(k_1, c = \omega/k_1) = 0$. Indeed, equations (7.104) and (7.105) yield

$$\omega^2 - c_T^2 \left(k_1^2 - \alpha^2 \right) = 0, \tag{7.111}$$

$$\omega^2 - \hat{c}_T^2 \left(k_1^2 + k_a\, \alpha \right) = 0. \tag{7.112}$$

The last equation provides the expression

$$\alpha = \left(\frac{c^2}{\hat{c}_T^2} - 1 \right) \frac{k_1^2}{k_a}, \tag{7.113}$$

with $\omega = c\, k_1$. Substituting from this into (7.111), we obtain

$$D(\omega, k_1) := \omega^2 - c_T^2 \left(k_1^2 - \frac{1}{k_a^2} \left(\frac{\omega^2}{\hat{c}_T^2} - k_1^2 \right)^2 \right) = 0. \tag{7.114}$$

This is not the most convenient form of the dispersion relation. Rather, accounting for $c = \omega/k_1$ the phase speed, and setting

$$s = \frac{c}{c_T}, \qquad K = \frac{k_1}{k_a}, \qquad \zeta = \frac{c_T}{\hat{c}_T}, \tag{7.115}$$

we can rewrite the dispersion relation as

$$\widetilde{D}(K, s) := K^2 - \frac{1 - s^2}{(1 - \zeta^2\, s^2)^2} = 0. \tag{7.116}$$

From this we see that s must be smaller than one, so that $c < c_T$. But with $\alpha > 0$, (7.113) imposes that $c > \hat{c}_T$, and we must also check that $\zeta > 1$, so that $\hat{c}_T < c_T$. Globally, the phase speed c is such that

$$\hat{c}_T < c < c_T. \tag{7.117}$$

This resembles the existence conditions for classical Love SAWs (slow layer of finite thickness superimposed on a semi-infinite substrate; See Section 7.3, Equation (7.58)). But while Love waves exhibit a series of modes, the present wave has only one mode given by (7.116) and satisfying (7.117). Nonetheless, the obtained mode is *dispersive* being characterized by the existence of the characteristic length scale $l_M = k_a^{-1}$ or wavelength $\lambda_M = 2\pi/k_a$.

7.4.3   *Canonical conservation laws for Murdoch linear SAWs*

These are the conservation laws that concern the whole physical system under consideration, hence all degrees of freedom **simultaneously**, in the present case both displacement fields in the half space $x_2 > 0$ (i.e., u_3) and at the material surface $x_2 = 0$ (i.e., $\hat{u}_3$). In nondissipative cases these equations are usually obtained by invoking the celebrated Noether's theorem which associates such a conservation law with each member of the invariance group of the Lagrangian (see Chapter 3). Here we simply use the naïve method which consists in performing simple manipulations on the already known local field equations, equations (7.104) and (7.105). With a view to obtaining the conservation equations of *energy* and *wave momentum*, this consists in multiplying these equations by the time derivative and the gradient of the relevant field and combining the results in the appropriate way.

A. Conservation of energy

Equations (7.105) and (7.104) are multiplied by $\hat{u}_{3,t}$ and $u_{3,t}$, respectively, with an obvious notation. After some manipulations, we obtain the two equations

$$\frac{\partial \widehat{H}}{\partial t} - \frac{\partial \widehat{Q}}{\partial x_1} - \mu\, u_{3,2}\, \hat{u}_{3,t} = 0, \qquad (7.118)$$

and

$$\frac{\partial H}{\partial t} - \frac{\partial Q_1}{\partial x_1} - \frac{\partial Q_2}{\partial x_2} = 0, \qquad x_2 > 0, \qquad (7.119)$$

where we have defined a "Hamiltonian" (total energy) density and an energy flux within the material surface $x_2 = 0$ by

$$\widehat{H} = \frac{1}{2}\left(\hat{\rho}_0\, \hat{u}_{3,t}^2 + \hat{\mu}\, \hat{u}_{3,1}^2\right), \qquad (7.120)$$

while for $x_2 > 0$ we have

$$H = \frac{1}{2} \left(\rho_0 \, \hat{u}_{3,t}^2 + \mu \, \Phi \right), \quad Q_1 = \mu \, u_{3,1} \, u_{3,t}, \quad Q_2 = \mu \, u_{3,2} \, u_{3,t}. \quad (7.121)$$

With all fields tending towards zero at infinity in depth, from (7.119) we deduce that

$$\frac{\partial}{\partial t} \int_0^{+\infty} H \, dx_2 - \frac{\partial}{\partial x_1} \int_0^{+\infty} Q_1 \, dx_2 + \mu \left(u_{3,2} \, u_{3,t} \right) \big|_{x_2=0} = 0. \quad (7.122)$$

On combining (7.118) and (7.122) while identifying $\hat{u}_{3,t}$ and $u_{3,t}$ at $x_2 = 0$, we obtain the *equation of conservation of energy* per unit thickness in the x_3-direction and unit length in the x_1 propagation direction in the form

$$\frac{\partial \widetilde{H}}{\partial t} - \frac{\partial \widetilde{Q}_1}{\partial x_1} = 0, \quad (7.123)$$

wherein

$$\widetilde{H} = \widehat{H} + \int_0^{+\infty} H \, dx_2, \quad \widetilde{Q}_1 = \widehat{Q} + \int_0^{+\infty} Q_1 \, dx_2. \quad (7.124)$$

B. Conservation of wave momentum

Instead of the previous manipulation we multiply equations (7.105) and (7.104) by $\hat{u}_{3,1}$ and $u_{3,1}$, respectively. We obtain thus

$$\frac{\partial \widehat{P}_1}{\partial t} + \frac{\partial \widehat{H}}{\partial x_1} + \mu \, u_{3,2} \, \hat{u}_{3,1} = 0, \quad (7.125)$$

and

$$\frac{\partial P_1}{\partial t} - \frac{\partial b_{11}}{\partial x_1} + \frac{\partial}{\partial x_2} (\mu \, u_{3,2} \, u_{3,1}) = 0, \quad (7.126)$$

where we defined the surface wave momentum component by

$$\widehat{P}_1 := -\hat{\rho}_0 \, \hat{u}_{3,t} \, \hat{u}_{3,1}, \quad (7.127)$$

and the bulk wave momentum by

$$P_1 := -\rho_0 \, u_{3,t} \, u_{3,1}, \quad b_{11} = -(L + \sigma_{31} \, u_{3,1}), \quad (7.128)$$

with

$$L = \frac{1}{2} \rho_0 \, u_{3,t}^2 - \frac{1}{2} \mu \, \Phi, \quad \sigma_{31} = \mu \, u_{3,1}. \quad (7.129)$$

But equation (7.125) presents an artefact due to the one-dimensionality of the considered system. In effect, we can note that in the present case

$$\widehat{H} = -\hat{b}_{11}, \quad \hat{b}_{11} = -(\widehat{L} + \hat{\sigma}_{31} \, u_{3,1}), \quad \widehat{L} = \frac{1}{2} \left(\hat{\rho}_0 \, \hat{u}_{3,t}^2 - \hat{\mu} \, \hat{u}_{3,1}^2 \right). \quad (7.130)$$

Here $\hat{b}_{11}$ is but one special component of the so-called *Eshelby material stress tensor*, where the second of (7.130) is its canonical definition — in 3D we have $b_{ji} = -(L\,\delta_{ji} + \sigma_{jk}\,u_{j,i})$ —, while the first of (7.130) is the announced artefact. Thus (7.125) reads

$$\frac{\partial \hat{P}_1}{\partial t} - \frac{\partial \hat{b}_{11}}{\partial x_1} + \mu\,u_{3,2}\,\hat{u}_{3,1} = 0. \tag{7.131}$$

In the same conditions as (7.122), by integration over depth we obtain

$$\frac{\partial}{\partial t}\int_0^{+\infty} P_1\,dx_2 - \frac{\partial}{\partial x_1}\int_0^{+\infty} b_{11}\,dx_2 - \mu\,u_{3,2}\,u_{3,1}\big|_{x_2=0} = 0. \tag{7.132}$$

On combining (7.125) and (7.132) at $x_2 = 0$, we obtain a *conservation equation for wave momentum* in total analogy with the energy conservation (7.123) as

$$\frac{\partial \tilde{P}_1}{\partial t} - \frac{\partial \tilde{b}_{11}}{\partial x_1} = 0, \tag{7.133}$$

with the definitions

$$\tilde{P}_1 = \hat{P}_1 + \int_0^{+\infty} P_1\,dx_2, \qquad \tilde{b}_{11} = \hat{b}_{11} + \int_0^{+\infty} b_{11}\,dx_2. \tag{7.134}$$

The strict conservation equations (7.123) and (7.133) hold true because our system is neither dissipative (no source term in (7.123)) nor materially inhomogeneous. However, we can well conceive a material inhomogeneity both for ρ_0 and μ in the depth direction x_2, i.e., some kind of continuous stratification. The general theory of material inhomogeneities is well equipped for that.

7.4.4 *Associated quasi-particle*

Following the methodology exploited in previous chapters and sections, the "point mechanics" of a so-called quasi-particle associated with the continuous wave motion is obtained by integrating the canonical conservation laws of wave momentum and energy over *a material volume element that is representative of this wave motion*. In the present case, integration over the depth coordinate x_2 in equations (7.123) and (7.133) is already effected. It remains to integrate these equations over one wavelength in the propagation direction at the material surface $x_2 = 0$. That is,

$$\frac{d}{dt}\int_0^{\lambda_M} \tilde{H}\,dx_1 = \left[\tilde{Q}\right]_0^{\lambda_M}, \tag{7.135}$$

and

$$\frac{d}{dt} \int_0^{\lambda_M} \widetilde{P} \, dx_1 = \left[\widetilde{b}_{11} \right]_0^{\lambda_M}. \tag{7.136}$$

On account of the real Murdoch wave solution, we can compute

$$\int_0^{+\infty} P \, dx_2 = \frac{\rho_0}{2\,\alpha} A^2 \, \omega \, k_1 \, \sin^2 \varphi,$$

$$\widetilde{P} = \left(\hat{\rho}_0 + \frac{\rho_0}{2\,\alpha} \right) A^2 \, \omega \, k_1 \, \sin^2 \varphi. \tag{7.137}$$

By integration of the second of these along x_1, this yields the momentum of the associated quasi-particle as

$$P_M = M_M \, c_M, \tag{7.138}$$

where $c = c_M = \omega/k_1$, $k_1 = k_M = 2\,\pi/\lambda_M$ and the "mass" M_M is defined by

$$M_M = \left(\hat{\rho}_0 + \frac{\rho_0}{2\,\alpha} \right) A^2 \, \pi \, k_M. \tag{7.139}$$

Like in previous cases this mass is proportional to the square of the wave amplitude and to a mass density. Here the involved mass density is a combination of the mass density of the material surface and of the substrate (the latter pondered by the depth behaviour). Note that this mass explodes with α going to zero, in effect when the material surface does not exist anymore and then does not allow for the existence of the considered SH wave mode. Furthermore, it depends on the wavelength as, contrary to some previously studied cases, we are facing here a *dispersive* wave system.

The *Newtonian motion* described by the resulting equation (7.80) will be *inertial* if and only if the right-hand side of this equation vanishes identically. But this is trivially satisfied because of the periodicity of the solution along x_1. Thus we have the inertial equation of motion

$$\frac{d}{dt} (M_M \, c_M) = 0, \tag{7.140}$$

where the point mass is given by (7.139) and c_M is the phase speed of the Murdoch surface wave.

The point mechanics of our quasi-particle is complete once we have the corresponding energy equation, because we recall that a "point mechanics" consists in a closed set of relations between a (rest) mass, a velocity, and a kinetic energy. This is obtained by evaluating the two sides of Equation (7.135) independently of the derivation of the motion equation (7.140). Again the right-hand side of (7.135) vanishes identically because of the periodicity of all fields. As to the left-hand side we first note that, after integration along x_2,

$$\int_0^{\lambda_M} \widetilde{H}\, dx_1 = \frac{A^2}{4}\, \lambda_M \left(\hat{\rho}_0\, \omega^2 + \frac{\rho_0}{2\,\alpha}\, \omega^2 + \hat{\mu}\, k_1^2 + \frac{\mu}{2\,\alpha}\, (k_1^2 + \alpha^2) \right). \quad (7.141)$$

Accounting now for the dispersion relation (7.55) and (7.56) this transforms to

$$\int_0^{\lambda_M} \widetilde{H}\, dx_1 = \frac{A^2}{4}\, \lambda_M \left(2 \left(\hat{\rho}_0 + \frac{\rho_0}{2\,\alpha} \right) \omega^2 + (\mu - \hat{\mu}\, k_a)\, \alpha \right). \quad (7.142)$$

But accounting for the definition of k_a, the last quantity within the parentheses vanishes identically leaving the result

$$E := \frac{1}{2} \int_0^{\lambda_M} \widetilde{H}\, dx_1 = \frac{1}{2}\, M_M\, c_M^2, \quad (7.143)$$

with a "mass" defined by (7.139) and c_M the phase velocity of the Murdoch SAW. Thus, with a vanishing right-hand side, Equation (7.135) yields the equation of conservation of energy of the associated quasi-particle in the Newtonian–Leibnizian form

$$\frac{d}{dt} E_M = \frac{d}{dt} \left(\frac{1}{2}\, M_M\, c_M^2 \right) = 0, \quad (7.144)$$

a natural form since there is no dissipation in the original wave system. Note finally that the energy of the quasi-particle appears as *purely kinetic* while originally two different kinetic energies and two different elastic energies were involved in the wave system.

7.4.5 *Consideration on the Lagrangian of the wave system*

With a Lagrangian given by equation (7.109), we envisage to evaluate the following quantity (average over one wavelength)

$$\langle L|_{x_2=0} \rangle_{\lambda_M} := \frac{1}{\lambda_M} \int_0^{\lambda_M} L|_{x_2=0}\, dx_1. \quad (7.145)$$

We want to prove that this is identically zero when the dispersion relation of Murdoch waves is satisfied for real solutions of the type

$$u_3(x_1, x_2; t) = A\, \exp(-\alpha\, x_2)\, \cos\varphi, \qquad \varphi = k_1\, x_1 - \omega\, t. \quad (7.146)$$

On account of straightforward calculations (*cf.* Rousseau and Maugin (2012b)), it is shown that

$$L|_{x_2=0} = \frac{1}{2}\, \hat{\rho}_0\, \hat{c}_T^2\, \alpha\, A^2\, k_a\, \sin^2\varphi - \frac{1}{4}\, \rho_0\, c_T^2\, A^2\, \alpha, \quad (7.147)$$

and, accounting for the definition of k_a,

$$\int_0^{\lambda_M} L|_{x_2=0} \, dx_1 = \frac{1}{2} \rho_0 \, c_T^2 \, \alpha \, A^2 \left(\int_0^{\lambda_M} \left(\sin^2 \varphi - \frac{1}{2} \right) dx_1 \right) \equiv 0. \quad (7.148)$$

This remarkable result, known for a one-dimensional harmonic-wave motion in the kinematic-wave theory (*cf.* Maugin (2007)), here is proved for a very special surface-wave motion (Murdoch wave) that is an *inhomogeneous* and *dispersive* wave motion (i.e., with a peculiar transverse or orthogonal behaviour). We already established such an identity in previous sections.

7.4.6   *Murdoch case as a limit of the Love case*

The setting for the existence of Murdoch surface waves obviously corresponds to a thin-film limit of the setting for the existence of Love waves. It is first checked that the Love dispersion relation (7.41) directly reduces to the Murdoch dispersion relation (7.114) which can be rewritten as

$$k^2 = k_a^2 \frac{\left(1 - \omega^2/\omega_T^2\right)}{\left(\omega^2/\widehat{\omega}_T^2 - 1\right)^2}, \quad (7.149)$$

where

$$\omega_T^2 = c_{T_2}^2 \, k^2 = c_T^2 \, k^2, \quad \widehat{\omega}_T^2 = c_{T_1}^2 \, k^2 = \hat{c}_T^2 \, k^2,$$

$$\hat{c}_T^2 = \frac{\hat{\mu}}{\rho}, \quad \hat{\mu} = \mu_1 \, d, \quad k_a = \frac{\mu}{\hat{\mu}}. \quad (7.150)$$

It is also checked from the expressions of mass, momentum and energy exhibited in Section 7.3 that we obtain the looked for reduction. Indeed, from Equation (7.92) one obtains the limit expression for the thin-film limit of the contribution from the superimposed layer (with d/λ going to zero)

$$M_1(\text{thin film}) = (\rho_1 \, d) \, \pi \, A^2 \, k, \quad (7.151)$$

and setting $\hat{\rho}_0 = \rho_1 d$ (density of mass per unit surface), $\rho_0 = \rho_2$, and $k_2 = \alpha$, we obtain from (7.83) the "mass" defined by (7.135). The same exact approximation can be obtained for the energy of the associated quasi-particle.

In conclusion of this section, we have unambiguously established the Newtonian–Leibnizian mechanics of the point-wise quasi-particle that can be associated with a linear Murdoch SAW by means of the exploitation of the corresponding canonical equations of conservation of energy and wave momentum. We also checked that its point mechanics corresponds to the thin-film limit of that of the quasi-particle associated with Love waves.

7.5   Conclusion

The above given developments show that it is possible to introduce the notion of guided quasi-particle in direct association with surface acoustic waves. Standard Rayleigh waves provided a perfect example for this possibility. Some generalization is also feasible as when an energy is granted to the limiting surface. If the latter of vanishing thickness is also given an inertia, then this property allows one to envisage the propagation of a pure SH, but dispersive, surface mode, with which one easily associates a quasi-particle in steady motion. The Love wave case is of interest in that it leads to an original point dynamics that involves both lower and upper bounds for its velocity, hence a scheme more complex than that of a standard Lorentzian–Einsteinian point particle The Murdoch case corresponds to the very thin plate hypothesis of Tiersten and this may be considered a limit of the configuration that favours the existence of Love surface waves. But now only one mode is involved, facilitating thus the analysis. This case allowed us to present all interesting features while avoiding the difficulties inherent in standard Love waves as exhibited in Section 7.3. Coupling with electric properties in electroelasticity will also favour the existence of mechanically pure SH waves in the next chapter. This will also offer us the possibility to examine the effects of the nonlinearity and viscosity of the substrate in Chapter 8.

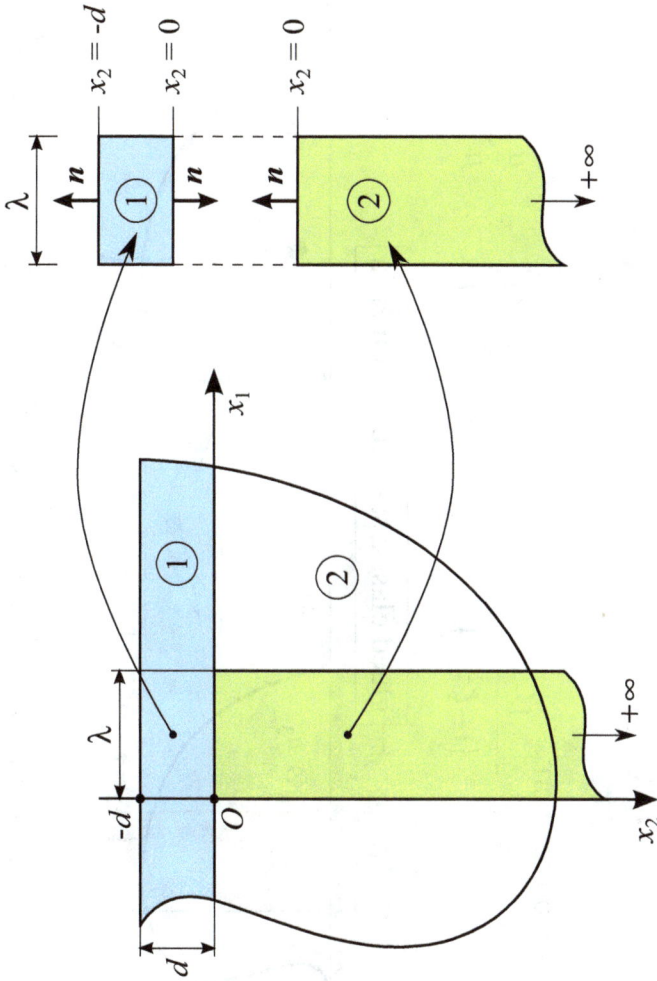

Plate 7.5   The case of Love SAWs.

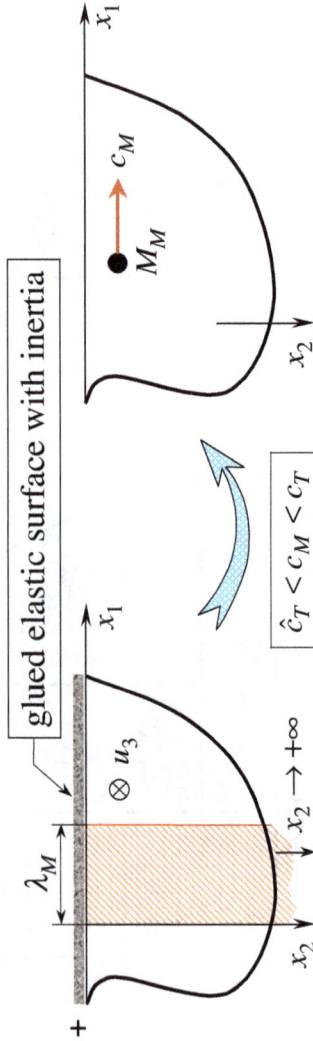

Plate 7.6   The case of SH Murdoch waves.

Chapter 8

Electroelastic surface waves in terms of quasi-particles

Object of the Chapter

This chapter considers the possible generalization of some results of Chapter 7 to the case of electroelastic surface waves. However, for the sake of simplicity in the analysis, a special illustrative mode of propagation is selected (so-called Bleustein–Gulyaev shear-horizontal waves) which allows one to associate with it a well-defined quasi-particle in inertial motion. Furthermore, this relative simplicity also permits us to outline the influence of a nonlinear elasticity of the substrate as well as that of a possible viscosity of this substrate. In the last case, the associated quasi-particle exhibits a non-inertial motion being driven by an effective friction force, while also moving no longer rigorously along the surface of the substrate. The presented treatment exploits the smallness of both nonlinearity and dissipation effects.

8.1   The notion of electroelastic surface wave

Electroelasticity is one of these coupled-field theories that puts under a single umbrella deformable processes in elastic solids and the effects of electric fields in such electrically polarisable materials. It is generally considered as taking place in the arena of crystal physics as its most common application is in the case of elastic crystals having the required degree of material *anisotropy*. This is classically illustrated by the case of so-called *piezoelectric crystals* that allow the existence of a *linear* electro-mechanical coupling due to the existence of a restricted symmetry (no centre of symmetry). Electroelasticity is a much larger field that includes nonlinear couplings and a variety of materials, including electrically polarisable, finitely deformable polymers, in which case even isotropy is allowed and the remarkable effect is none other than *electrostriction* (a produced strain quadratic in the electric field). We refer the reader to Eringen and Maugin (1990a) and the book of Maugin (1988) for a general formulation in the framework of finite-strain continuum thermo-mechanics. Many nonlinear dynamical effects and their applications are treated in the monograph of Maugin *et al.* (1992b). In the present chapter we shall focus attention on the case of piezoelectricity for which the basic equations will be recalled in Section 8.2.

Surface waves of a mixed electro-elastic nature are of particular interest because of their importance in the applied field of signal processing. The reason for this is that due to their short wavelength — compared to that of electromagnetic signals — they afford the conception of some very useful instruments or technical devices in a very limited space. These include devices such as transducers, dynamic condensers, convolvers and correlators of signals, *etc.*

In the case of interest here, the general framework reduces to *quasi-electrostatics*, by which it is understood that only the electrostatic part of Maxwell's equations intervenes in the problem while the motion and de-formation remain dynamical — and the electric field still varies with time —, but obviously limited to a reasonable range of velocities and frequencies. As a consequence, the electroelastic surface waves to be considered materialize a dynamic coupling between small strains and the time-varying electrostatic potential. It is this drastic simplification that allows for a relatively simple treatment although the field, in a sense, has become four-dimensional (three components of the elastic displacement and one scalar of electric nature). The required anisotropy generally leads to the existence of so-called *generalized Rayleigh waves* (with an elastic polarization not restricted to the sagittal plane). This matter is discussed in Dieulesaint and Royer (2000). For our purpose, however, we shall select a material symmetry that allows a decoupling between an essentially Rayleigh type of mode (polarized in the sagittal plane) and an elastic component represent-ing an SH (for shear horizontal) mode. Again, this simplification allows us to easily introduce the notion of quasi-particle (see Sections 8.4 and 8.5) as-sociated with this type of surface waves (introduced in Section 8.3), which type was invented by Bleustein (1968) and Gulyaev (1969) in remarkably short papers. Of course, the electric boundary conditions at the top surface of a semi-infinite substrate here play a fundamental role. They require special attention as it may happen that an electric field is also present out-side the material region. But still it is again the relative simplicity of the mechanical description that allows us to envisage perturbations by a *weak elastic nonlinearity* of the substrate (see Section 8.5) as also by some *viscosity* that the substrate may present (see Section 8.6). The contents of this chapter are essentially based on papers by the authors (Maugin and Rousseau (2010b); Rousseau and Maugin (2012a,c)).

8.2 Basic equations of piezoelectricity

To be sufficiently general, we consider the electroelasticty of homogeneous materials whose material symmetry admits the phenomenon of piezoelectricity (general formulations to be found in books, e.g., Maugin (1988); Eringen and Maugin (1990a)). Only small strains and quasi-electrostatics are envisaged. Let $\boldsymbol{\sigma}$, $\mathbf{e}$, $\mathbf{u}$, $\mathbf{D}$, $\mathbf{E}$, $\mathbf{P}$, and ϕ denotes the (symmetric) Cauchy stress, the small strain, the elastic displacement, the electric displacement, the electric field, the electric polarization per unit volume, and the quasi-static electric potential. In any regular material region B the two electromechanical *field equations* are

$$\frac{\partial \mathbf{p}}{\partial t} - div\,\boldsymbol{\sigma} = 0, \qquad \nabla \cdot \mathbf{D} = 0, \tag{8.1}$$

where $\mathbf{p} = \rho_0\,\mathbf{v}$. Here we have no externally applied force, no ponderomotive electromagnetic force — by virtue of the linearity of the theory — and no bulk free electric charge (we consider a dielectric medium). The Cauchy stress $\boldsymbol{\sigma}$ is symmetric in the considered approximation; ρ_0 is a constant prescribed matter density. The material is considered homogeneous, but possibly of anisotropy class allowing for a piezoelectric coupling (linear electroelastic coupling, no centre of symmetry). Here,

$$\mathbf{e} = (\nabla \mathbf{u})_S, \qquad \mathbf{v} = \frac{\partial \mathbf{u}}{\partial t}, \tag{8.2}$$

and

$$\mathbf{D} = \varepsilon_0\,\mathbf{E} + \mathbf{P}, \qquad \mathbf{E} = -\nabla\phi. \tag{8.3}$$

The last equation follows from the reduced form of Faraday's equation, $\nabla \times \mathbf{E} = \mathbf{0}$, in quasi-electrostatics; ε_0 is the vacuum electric permeability.

A typical Lagrangian density per unit volume for the present nondissipative theory is given by

$$L = \frac{1}{2}\,\rho_0\,\mathbf{v}^2 + \frac{1}{2}\,\varepsilon_0\,\mathbf{E}^2 - W(\mathbf{e}, \mathbf{E}), \tag{8.4}$$

where W is the material electromechanical interaction energy.

The mechanical linear momentum and the constitutive equations are such that

$$\mathbf{p} = \frac{\partial L}{\partial \mathbf{v}}, \quad \boldsymbol{\sigma} = -\frac{\partial L}{\partial \mathbf{e}} = \frac{\partial W}{\partial \mathbf{e}}, \quad \mathbf{D} = \frac{\partial L}{\partial \mathbf{E}} = \varepsilon_0 \mathbf{E} + \mathbf{P}, \quad \mathbf{P} = -\frac{\partial W}{\partial \mathbf{E}}. \quad (8.5)$$

For a linear theory, the expression of the energy density W is quadratic, that is, explicitly in Cartesian tensor components,

$$W(\mathbf{e}, \mathbf{E}) = \frac{1}{2} C_{ijkl}\, e_{ij}\, e_{kl} - e_{qij}\, E_q\, e_{ij} - \frac{1}{2} \chi_{ij}\, E_i\, E_j, \quad (8.6)$$

with symmetries

$$C_{(ij)(kl)} = C_{klij}, \qquad e_{qij} = e_{qji}, \qquad \chi_{ij} = \chi_{ji}, \quad (8.7)$$

for the tensorial coefficients of elasticity, piezoelectricity and electric susceptibility. Accordingly, the constitutive equations read:

$$\sigma_{ij} = C_{ijkl}\, e_{kl} - e_{qij}\, E_q = C_{ijkl}\, u_{k,l} + e_{qij}\, \phi_{,q}, \quad (8.8)$$

$$D_i = \varepsilon_{ij}\, E_j + e_{ipq}\, e_{pq} = -\varepsilon_{ij}\, \phi_{,j} + e_{ipq}\, u_{p,q}, \quad (8.9)$$

$$\varepsilon_{ij} = \varepsilon_0\, \delta_{ij} + \chi_{ij} = \varepsilon_{ji}.$$

These are to be substituted in the field equations (8.1) and relevant boundary conditions. The latter are of mechanical and electric types.

Mechanical boundary conditions at the regular boundary ∂B of B, $\partial B = \partial B_\sigma \cup \partial B_u$:

- Free boundary:

$$\mathbf{n} \cdot \boldsymbol{\sigma} = \mathbf{0} \quad \text{at } \partial B_\sigma; \quad (8.10)$$

- Imposed displacement:

$$\mathbf{u} = \mathbf{u}_0 \quad \text{at } \partial B_u; \quad (8.11)$$

Electric boundary conditions. They are of two types ($\partial B = \partial B_D \cup \partial B_\phi$):

(a) Matching to an external electric field (the symbol [..] indicates the jump of is enclosure):

- No density of charges at ∂B_D

$$\mathbf{n} \cdot [\mathbf{D}] = 0 \quad \text{at } \partial B_D \quad (8.12)$$

- No jump in ϕ:

$$[\phi] = 0; \quad (8.13)$$

(b) Grounding with an electrode at ∂B_ϕ:

- Fixed potential:

$$\phi = \phi_0 \quad \text{say } 0; \quad (8.14)$$

- Prescribed surface density of electric charges:

$$\mathbf{n} \cdot \mathbf{D} = w. \quad (8.15)$$

8.3 Conservation laws of energy and wave momentum in electroelasticity

In contrast to the *field equations* (8.1) given for each field separately, we have seen in Chapter 3 that in field theory we also have *conservation laws* that apply to the *whole* physical system under consideration. This is trivially the case of the *energy equation* which here applies to the combined system of mechanical and electric quantities. In the absence of dissipation, this local conservation law here reads (*cf.* Maugin (1988))

$$\frac{\partial}{\partial t} H - \frac{\partial}{\partial x_j} \left(\sigma_{ji} \, \dot{u}_i + D_j \, \dot{\phi} \right) = 0, \tag{8.16}$$

where the Hamiltonian density is given by

$$H = \mathbf{p} \cdot \mathbf{v} - L = \frac{1}{2} \rho_0 \, \dot{\mathbf{u}}^2 + \overline{W}, \qquad \overline{W} = W - \frac{1}{2} \varepsilon_0 \, \mathbf{E}^2. \tag{8.17}$$

In a naïve way this can be constructed by taking the inner product of the first of (8.16) by $\dot{\mathbf{u}}$ and combining with the equation obtained by multiplying the second of (8.16) by $\dot{\phi}$ and accounting for (8.5). Of course, this can also be obtained by application of Noether's theorem to a variational formulation based on the Lagrangian densisty (8.4) for invariance under time translations since the considered system is not dissipative.

In the like manner, the *conservation equation of wave momentum* can be obtained by multiplying the first and second of equations (8.1) by gradients of the fields, i.e., $\nabla \mathbf{u}$ and $\nabla \phi$, respectively. After some manipulations accounting for (8.5), a computation let to the reader by way of exercise will produce the looked conservation law in the form of the following *vectorial* equation whose Cartesian components are

$$\frac{\partial}{\partial t} p_i^w - \frac{\partial}{\partial x_j} b_{ji} = 0, \tag{8.18}$$

where we have defined the following quantities:

$$p_i^w := -\rho_0 \, \dot{u}_j \, u_{j,i}, \qquad b_{ji} = -(L \, \delta_{ji} + \sigma_{jk} \, u_{k,i} + D_j \, \phi_{,i}),$$
$$L = \frac{1}{2} \rho_0 \, \dot{\mathbf{u}}^2 - \overline{W}, \tag{8.19}$$

called respectively the wave momentum, the so-called Eshelby (non symmetric) stress tensor, and an effective Lagrangian density. The application of Noether's theorem would have led to equation (8.18) by considering the total physical system invariant under material space translations reflecting material homogeneity. Of course we note the canonical expression of the

Eshelby stress in which mechanical and electric fields play a fully parallel role albeit of differing tensorial order. As a matter of fact, the second of (8.19) is none other than

$$\mathbf{b} = -\left(L\mathbf{1} - \frac{\partial L}{\partial(\nabla\mathbf{u})} \cdot (\nabla\mathbf{u})^T - \frac{\partial L}{\partial(\nabla\phi)} \otimes \nabla\phi\right). \tag{8.20}$$

The reader may wonder what happens when elementary dissipative processes are associated with the mechanical and electric properties, respectively viscosity and electric relaxation. The strategy of Noether' theorem no longer applies, but one can still mimic the above sketched derivation. This is achieved by multiplication of Equations (8.16) by the appropriate time rates and field gradients and assuming that stress now includes a viscous contribution in addition to a part derivable from the energy potential, while the electric polarization itself has a relaxation part added to the one derivable from the potential. Indeed, as shown by Rousseau and Maugin (2011b), it is found that equations (8.16) and (8.18) are consistently modified to read (note that $\boldsymbol{\sigma}$ and $\mathbf{D}$ now include both energy-derived and dissipative contributions):

$$\begin{aligned}
&\frac{\partial}{\partial t}\left(\frac{1}{2}\rho_0\,\dot{\mathbf{u}}^2 + \overline{W}\right) - \frac{\partial}{\partial x_j}\left(\sigma_{ji}\,\dot{u}_i + D_j\,\dot{\phi}\right) \\
&= -\left(\sigma_{ji}^{\text{visco}}\,\dot{u}_{i,j} + P_j^{\text{relax}}\,\dot{\phi}_{,j}\right),
\end{aligned} \tag{8.21}$$

and

$$\frac{\partial}{\partial t}p_i^w - \frac{\partial}{\partial x_j}b_{ji} = -\left(\sigma_{jk}^{\text{visco}}\,u_{k,ji} + P_j^{\text{relax}}\,\phi_{,ji}\right), \tag{8.22}$$

the expressions (now no more than a mere definition) of H and L being formally unchanged although no variational principle is any longer implemented. This was established in a general theory of dissipative processes in deformable continua by one of the authors (Maugin (2006)).

8.4 The Bleustein–Gulyaev surface wave

8.4.1 *The general surface wave problem in piezoelectric materials*

If we consider the general problem of surface wave propagation with a substrate made of an anisotropic piezoelectric elastic crystal and a limiting top surface $x_2 = 0$ (see the general notation in Chapter 7), then we will have to account for the following equations:

- For $x_2 > 0$:

$$C_{ijkl}\, u_{k,lj} + e_{kij}\, \phi_{,kj} = \rho_0\, \ddot{u}_i, \tag{8.23}$$

and

$$\varepsilon_{ij}\, \phi_{,ji} - e_{jkl}\, u_{k,lj} = 0; \tag{8.24}$$

- at $x_2 = 0$ (mechanically free surface):

$$(C_{ijkl}\, u_{k,l} + e_{kij}\, \phi_{,k})\, n_j = 0, \tag{8.25}$$

$$\mathbf{n} \cdot [\mathbf{D}] = 0, \tag{8.26}$$

$$\phi = \phi_0 \ \text{(potential prescribed at } x_2 = 0, \text{ e.g., ground)}, \tag{8.27}$$

$$[\phi] = 0 \ \text{(continuity then (8.26) applies)}; \tag{8.28}$$

with

$$\phi \to 0 \ \text{as} \ x_2 \to \pm\infty, \tag{8.29}$$

$$|\mathbf{u}| \to 0 \ \text{as} \ x_2 \to +\infty. \tag{8.30}$$

The general system involves four coupled amplitudes so that the Rayleigh wave problem has a general solution of the form

$$\left\{ u_l(x_1, x_2, t), \ l = 1, 2, 3, 4 \right\} = \left\{ u_i, \ i = 1, 2, 3; \ u_4 = \phi \right\}, \tag{8.31}$$

with

$$u_l = u_l^0 \exp(-k_1\, \chi\, x_2) \exp\left(i\, (\omega\, t - k_1\, x_1) \right), \qquad \Re(\chi) > 0. \tag{8.32}$$

The general solution for $x_2 > 0$ that must satisfy the boundary conditions is written as the four Cartesian vector

$$u_l = \sum_\alpha A_\alpha\, u_l^{0(\alpha)} \exp(-k_1\, \chi_\alpha\, x_2) \exp\left(i\, (\omega\, t - k_1\, x_1) \right), \tag{8.33}$$

where the coefficients A_α are determined by the boundary conditions. The mechanical boundary condition (8.25) yields

$$\sum_{\alpha=1}^{3} a_i^{(\alpha)}\, A_\alpha = -a_i^{(4)}\, A_4, \quad i = 1, 2, 3, \tag{8.34}$$

where

$$a_i^{(\alpha)} := (i\, C_{i21l} + \chi_\alpha\, C_{i22l})\, u_l^{0(\alpha)} + i\, e_{1i2} + \chi_\alpha\, e_{2i2}. \tag{8.35}$$

If $A_4 = 0$, then the propagation velocity of the surface mode results from the compatibility condition for solving the *homogeneous* linear system (8.34) with $e_{ijk} = 0$. If the substrate is piezoelectric, then equations (8.34) provide the coefficients A_i, $i = 1, 2, 3$ in function of A_4. The propagation velocity of the surface mode will then be determined by the relevant electric boundary conditions. The problem rapidly becomes very technical. We refer to Dieulesaint and Royer (2000) and Coquin and Tiersten (1967) for details. The resulting surface waves are of interest in the design of transducers, delay lines, and wave filters.

We witness the complexity of the general case for which the three elastic displacement components and the electric potential are coupled for $x_2 > 0$. The corresponding Rayleigh surface waves carry three components of elastic displacement and are accompanied by an electric field polarized in the sagittal plane Π_S since $E_3 = -\partial\phi/\partial x_3 = 0$. It may happen for certain orientations of the sagittal plane with respect to the symmetry axes of the crystal that the four-dimensional system of amplitudes split in two systems having only two or three mechanical or electric components. This is indeed the case for the so-called Bleustein–Gulyaev (for short BG) waves. These were discovered for a specific crystal symmetry (symmetry axis of order six orthogonal to Π_S) by Bleustein (1968) and Gulyaev (1969) — Russian original also in 1968 — providing, apart form a standard Rayleigh system (that we shall not discuss), a pure mechanical SH SAW, however coupled to the electric potential ϕ. We shall deal at some length with this case since it provides a basis for all subsequent sections in this chapter.

8.4.2   *The Bleustein–Gulyaev surface wave problem per se*

The setting is the one indicated in the coloured Plate 8.1. The material symmetry chosen is that of an elastic crystal of symmetry axis of order six (e.g., 6mm class) orthogonally to the sagittal plane (see Maugin (1988, pp. 250–254)). This allows for the existence of a pure shear horizontal (SH) elastic motion of component u_3 that however remains coupled to the potential ϕ. The surviving components of the constitutive equations (8.8)–(8.9) read

$$\sigma_{23} = c_{44}\, u_{3,2} + e_{15}\, \phi_{,2}, \qquad \sigma_{13} = c_{44}\, u_{3,1} + e_{15}\, \phi_{,1}, \qquad (8.36)$$

$$D_1 = e_{15}\, u_{3,1} - \varepsilon_{11}\, \phi_{,1}, \qquad D_2 = e_{15}\, u_{3,2} - \varepsilon_{11}\, \phi_{,2}. \qquad (8.37)$$

For $x_2 > 0$, the surviving field equations (8.1) are given by

$$c_{44}\, \boldsymbol{\nabla}^2 u_3 + e_{15}\, \boldsymbol{\nabla}^2 \phi = \rho_0\, \frac{\partial^2 u_3}{\partial t^2}, \qquad e_{15}\, \boldsymbol{\nabla}^2 u_3 - \varepsilon_{11}\, \boldsymbol{\nabla}^2 \phi = 0, \qquad (8.38)$$

where ∇^2 is the Laplacian in (x_1, x_2), and c_{44}, e_{15}, and ε_{11} are the only elastic, piezoelectric and electric material coefficients involved written in the Voigt notation. From the second of (8.38) we can introduce an effective electric scalar potential ψ such that

$$\psi = \phi - \frac{e_{15}}{\varepsilon_{11}} u_3. \tag{8.39}$$

Equations (8.38) then take the form

$$\bar{c}_T^2 \, \nabla^2 u_3 = \frac{\partial^2 u_3}{\partial t^2}, \qquad \nabla^2 \psi = 0. \tag{8.40}$$

Here, we have defined the piezoelectrically stiffened elastic squared speed $\bar{c}_T^2$, the piezoelectrically stiffened shear elastic coefficient $\bar{c}_{44}$, and the electromechanical coupling factor K by

$$\bar{c}_T^2 = \frac{\bar{c}_{44}}{\rho_0}, \qquad \bar{c}_{44} = c_{44}\left(1 + K^2\right), \qquad K^2 = \frac{e_{15}^2}{\varepsilon_{11}\, c_{44}}. \tag{8.41}$$

The Bleustein–Gulyaev surface wave solution of equations (8.40) in the positive half plane $x_2 > 0$ *a priori* reads

$$\begin{aligned} u_3 &= A \, \exp(-k_1 \chi \, x_2) \cos(k_1 \, x_1 - \omega \, t), \\ \psi &= B \, \exp(-k_1 \chi \, x_2) \cos(k_1 \, x_1 - \omega \, t). \end{aligned} \tag{8.42}$$

Substitution of the first of these in $(8.40)_1$ — a standard wave equation in 2D — yields the "dispersion relation"

$$D(\omega, k_1, \chi) = \omega^2 - \bar{c}_T^2 \, k_1^2 \left(1 - \chi^2\right) = 0, \tag{8.43}$$

while the second of (8.42) substituted for in the second of (8.40) — just a Laplace equation in 2D — of necessity yields $\chi = 1$ for the potential ψ, so that $(8.42)_2$ should be written as

$$\psi = B \, \exp(-k_1 x_2) \cos(k_1 \, x_1 - \omega \, t). \tag{8.44}$$

The true dispersion relation $D_{BG}(\omega, k_1) = 0$ will be obtained by eliminating $k_2 = k_1 \chi$ from (8.43). This is achieved thanks to the boundary conditions (BC) at the limiting plane $x_2 = 0$. Here we first consider the case of a *mechanically free, electrically grounded surface* $x_2 = 0$. The relevant boundary conditions are given by (8.10) — vanishing of stress component $\sigma_{23} = \sigma_{32}$ — and (8.14) with $\phi_0 = 0$. On account of $(8.42)_1$, (8.44), and (8.39), these BCs yield the following system for the amplitudes A and B:

$$\bar{c}_{44} \, k_1 \, \chi \, A + e_{15} \, k_1 \, B = 0, \qquad e_{15} \, A + \varepsilon_{11} \, B = 0. \tag{8.45}$$

This admits a non-zero solution for

$$\chi = \overline{K}^2; \qquad \overline{K}^2 := \frac{e_{15}^2}{\varepsilon_{11}\, \bar{c}_{44}} = \frac{K^2}{1 + K^2}, \tag{8.46}$$

while

$$B = -\frac{e_{15}}{\varepsilon_{11}} A. \tag{8.47}$$

Hence (8.43) yields the true "dispersion" relation for Bleustein–Gulyaev surface waves (for a grounded boundary)

$$D_{GB}(\omega, k_1) = \omega^2 - c_{BG}^2 k_1^2 = 0; \qquad c_{BG}^2 = \bar{c}_T^2 \left(1 - \overline{K}^4\right). \tag{8.48}$$

Note for further use that the Lagrangian density (8.4) reduces to the following expression:

$$L = \frac{1}{2} \rho_0 \left(\frac{\partial u_3}{\partial t}\right)^2 + \frac{1}{2} \varepsilon_{11} \left(\phi_{,1}^2 + \phi_{,2}^2\right) - \frac{1}{2} c_{44} \left(u_{3,2}^2 + u_{3,1}^2\right)$$
$$+ e_{15} \left(u_{3,2}\,\phi_{,2} + u_{3,1}\,\phi_{,1}\right), \tag{8.49}$$

where eventually ϕ will have to be replaced by its expression in terms of ψ and $u_{3,2}$. As a matter of fact, it is immediately noticed that equations (8.40) are derivable from the following Lagrangian density for $x_2 > 0$:

$$L = \frac{1}{2} \rho_0 \left(\frac{\partial u_3}{\partial t}\right)^2 + \frac{1}{2} \varepsilon_{11} \left(\psi_{,1}^2 + \psi_{,2}^2\right) - \frac{1}{2} \bar{c}_{44} \left(u_{3,2}^2 + u_{3,1}^2\right), \tag{8.50}$$

what formally amounts to replacing the symbols ϕ and c_{44} in (8.49) by the symbols ψ and $\bar{c}_{44}$, and discarding the coupling. We also need to record that D_2 is now given by

$$D_2 = -\varepsilon_{11}\,\psi_{,2}, \quad \text{for } x_2 > 0. \tag{8.51}$$

8.4.3 *Dynamics of the associated quasi-particle*

A. *Preliminary remark*

Following along the path of previous chapters, we have to integrate Equation (8.18) over a volume representative of the BG wave motion. This volume will be of one wave length in the propagation direction, one unit in the thickness direction x_3 and in depth from the surface $x_2 = 0$ to infinity, thus:

$$\Omega = [x_{10}, x_{10} + \lambda^w] \times [0, +\infty) \times [0, 1], \tag{8.52}$$

where λ^w is one wave length chosen at any place $x_{10} = (x_1)_0$ along the path of the wave.

Let us examine what happens at the surface boundary $x_2 = 0$. The problem being independent of x_3 and the outward normal pointing in the direction of negative x_2's, we have to consider the quantities $(\mathbf{n} \cdot \mathbf{b})_1 = -b_{21}$

and $(\mathbf{n} \cdot \mathbf{b})_2 = -b_{22}$. With boundary conditions (8.10) and (8.14), the second of (8.19) provides the quantities

$$- b_{21} = 0, \quad -b_{22} = L \quad \text{at } x_2 = 0. \tag{8.53}$$

But the reduced Lagrangian is given by (8.50) for the present solution. Now insert surface-wave solutions in $(8.53)_2$. There will appear only terms in $\sin^2 \varphi$ and $\cos^2 \varphi$ with the phase $\varphi = k_1 x_1 - \omega t$. That is, taking the average over a wavelength $\lambda = 2\pi/k_1$ along x_1 at $x_2 = 0$, we obtain

$$- \langle b_{22} \rangle \propto \frac{1}{2} \rho_0 A^2 \omega^2 - \frac{1}{2} \bar{c}_T^2 k_1^2 (1 - \chi^2) A^2. \tag{8.54}$$

Accounting now for the relation $(8.44)_1$, by virtue of (8.48) we obtain the following remarkable result:

$$\langle -\mathbf{n} \cdot \mathbf{b} \cdot \mathbf{e}_2 \rangle \big|_{x_2=0} \propto \frac{1}{2} \rho_0 A^2 D_{GB} \equiv 0, \tag{8.55}$$

where the average is one taken along the propagation direction over one wavelength, and D_{GB} is none other than the "dispersion" relation for Bleustein–Gulyaev SAWs.

In the present case, we note that $(8.52)_1$ ($b_{21} = 0$) is valid point wise at any point in the propagation wave-length interval while only the *average* over this interval of b_{22} vanishes. This is important in the sequel.

B. Momentum of the quasi-particle

Now we perform the integration of (8.18) over the volume (8.52), accounting for the result (8.55), i.e., (see coloured Plate 8.1 with $U = A$)

$$\frac{d}{dt} \int_\Omega \mathbf{p}^w \, dV = \int_\Omega \operatorname{div} \mathbf{b} \, dV = \int_{\partial\Omega} \mathbf{n} \cdot \mathbf{b} \, dS, \tag{8.56}$$

where $\mathbf{n}$ is the outward pointing unit normal. We have to evaluate the x_1 and x_2 components of this equation, hence p_1^w, p_2^w and $(\mathbf{n} \cdot \mathbf{b})_1$ and $(\mathbf{n} \cdot \mathbf{b})_2$, making abstraction of the unit thickness, i.e., as if (8.56) was evaluated in the sagittal plane. First we note that

$$P_2^w = \int \left(- \rho_0 \dot{u}_3 u_{3,2} \right) dV = \int_0^{+\infty} \int_{x_0}^{x_0+\lambda} \left(- \rho_0 \dot{u}_3 u_{3,2} \right) dx_1 \, dx_2. \tag{8.57}$$

The integral along x_1 is equivalent to taking the average over a wavelength at a fixed x_2. But this vanishes for the considered SAW because the term within parentheses involves a product term in $\sin \varphi \times \cos \varphi$. The other component of (8.56) is more interesting. For this, replace index 2 in both sides of (8.57) by index 1. Accounting for the average along x_1 this yields

$$
\begin{aligned}
P_1^w &= \int \left(-\rho_0 \, \dot{u}_3 \, u_{3,1} \right) dV \\
&= -\int_0^{+\infty} \frac{\pi}{2 \, k_1} \rho_0 \, \omega \, A^2 \, k_1 \, \exp(-k_1 \, \chi \, x_2) \, dx_2 = \frac{\pi}{2 \, k_2} \rho_0 \, \omega \, A^2,
\end{aligned}
\tag{8.58}
$$

or

$$
P_1^w = \left(\frac{\rho_0 \, A^2}{2 \, k_1} \right) \left(\frac{\pi \, \omega}{\overline{K}^2} \right),
\tag{8.59}
$$

on account of (8.48). The first parenthesis in the right-hand side of this equation is a mass $(kg/m^3) \times m^2 \times$ m. The second parenthesis is the inverse of a time, but the whole thing was computed as if per unit surface, so that the true physical dimension of P_1^w is a momentum (mass $\times$ velocity). Finally, re-introducing c_{BG}, we have

$$
P_1^w = M\left(A; \overline{K} \right) c_{BG}, \qquad M = M_{BG} := \frac{\rho_0 \, \pi \, A^2}{2 \, \overline{K}^2},
\tag{8.60}
$$

per unit thickness. The mass M apparently blows up as the piezoelectricity coefficient goes to zero at a fixed amplitude. But this amplitude also goes to zero as the wave cannot exist as a surface wave in the absence of coupling.

It remains to evaluate the right-hand side of Equation (8.56). As x_2 goes to infinity all relevant fields tend to zero and the extreme bottom part of the domain will not contribute any surface term. This also applies jointly to the contribution along the lateral sides L and R of the band because of the periodicity along x_1 and the opposite sign of the two external unit normals. Finally, the relevant terms at the surface $x_2 = 0$ are evaluated thanks to the first of (8.53) and (8.54). Summing up, the x_2-component of (8.56) yields zero on both sides, while the remaining x_1-component has a vanishing right-hand side, and there remains the left-hand side, so that the surviving global equation of wave momentum *in the propagation direction* reads

$$
\frac{d}{dt} \left(M\left(A; \overline{K} \right) c_{BG} \right) = 0.
\tag{8.61}
$$

This can be interpreted as the *inertial motion of a quasi-particle* of mass M_{BG} in a 1D- straight motion — guided by the top surface — with constant velocity solution of the unperturbed Bleustein–Gulyaev problem. Equation (8.61) is a Newtonian equation of motion for a "particle". Its "mass" is a measure of the amplitude of the signal. It is "informative" of the solution. We can check by computation that the "energy" of the obtained quasi-particle appears as purely "kinetic" and obeys Newtonian mechanics although it does include the effects of various [kinetic, potential (i.e., elastic), electrostatic, and electromechanical (here, piezoelectric)] origins in the continuum description. Indeed, the energy density associated with the Lagrangian density (8.50) reads $E = K + W$ (where K here is the kinetic energy), i.e.

$$E = \frac{1}{2}\rho_0 \left(\frac{\partial u_3}{\partial t}\right)^2 - \frac{1}{2}\varepsilon_{11}\left(\psi_{,1}^2 + \psi_{,2}^2\right) + \frac{1}{2}\bar{c}_{44}\left(u_{3,2}^2 + u_{3,1}^2\right). \tag{8.62}$$

Proceeding just like before with the already considered semi-infinite vertical band of the sagittal plane, we account for the solutions (8.42) and (8.44), and for (8.46) that results from the boundary conditions, and obtain, avoiding a few steps in the proof,

$$E_{qp} = \frac{\rho_0\,\pi\,A^2}{8\,\overline{K}^2}\left(c_{BG}^2 + \frac{\bar{c}_{44}}{\rho}\left(1 - \frac{e_{15}^2}{\varepsilon_{11}\,\bar{c}_{44}}\right)\left(1 + \overline{K}^2\right)\right), \tag{8.63}$$

or

$$\begin{aligned}
E_{qp} &= \frac{\rho_0\,\pi\,A^2}{8\,\overline{K}^2}\left(c_{BG}^2 + \frac{\bar{c}_{44}}{\rho}\left(1 - \overline{K}^2\right)\left(1 + \overline{K}^2\right)\right)\\
&= \frac{\rho_0\,\pi\,A^2}{8\,\overline{K}^2}\left(c_{BG}^2 + c_{BG}^2\right).
\end{aligned} \tag{8.64}$$

That is, with M_{BG} given by the second of (8.60),

$$E_{qp}(BG) = \frac{1}{2}\,M_{BG}\,c_{BG}^2. \tag{8.65}$$

Q.E.D. We remind the reader that this is per unit thickness in the x_3 direction.

8.4.4   *Another case of electric boundary condition*

If the electric condition (8.13) prevails, then matching with an external electric solution $\phi = C\,\exp(k_2\,x_2)\,\cos(k_1\,x_1 - \omega t)$ for $x_2 < 0$ applies at $x_2 = 0$. We have to consider the electrostatic equation $\nabla^2\phi = 0$ for $x_2 < 0$, so that $k_2 = k_1$ or $\chi = 1$ for electrostatics in the region $x_2 < 0$ (See coloured Plate 8.2). Since u_3 is inexistent for $x_2 < 0$, we can as well write ψ instead of ϕ, so that in we can write the solution

$$\psi = C\,\exp(k_1\,x_2)\,\cos(k_1\,x_1 - \omega t) \quad \text{for } x_2 < 0. \tag{8.66}$$

At $x_2 = 0$, the field ϕ is to be continuous (*cf.* equation (8.13)), so that

$$C = B - c_{44}\,K^2\,A. \tag{8.67}$$

We still need to satisfy the no-stress condition and the continuity of the normal component of $\mathbf{D}$, Equation (8.12). The latter yields

$$+ \varepsilon_0 \left(\phi_{,2}\right)^{\text{out}} - \left(e_{15}\, u_{3,2} - \varepsilon_{11} \left(\phi_{,2}\right)^{\text{in}}\right) = 0. \tag{8.68}$$

By the continuity of ϕ at $x_2 = 0$ and for the Bleustein–Gulyaev (u_3, ϕ) solution, this gives

$$e_{15}\, A + \left(\varepsilon_0 + \varepsilon_{11}\right) B = 0. \tag{8.69}$$

This is the only change in the problem formulation. Hence,

$$k_2 = \widetilde{K}^2 k_1, \qquad B = -\frac{e_{15}}{\varepsilon_0 + \varepsilon_{11}}\, A; \qquad \widetilde{K}^2 \equiv \frac{\overline{K}^2}{\varepsilon_0 + \varepsilon_{11}}. \tag{8.70}$$

It follows that (8.48) is replaced by

$$D_{GB}(\omega, k_1) = \widetilde{D}_{BG}(\omega, k_1) := \omega^2 - \tilde{c}_{BG}^2\, k_1^2 = 0;$$
$$\tilde{c}_{BG}^2 = \bar{c}_T^2 \left(1 - \widetilde{K}^4\right). \tag{8.71}$$

Since $\widetilde{K}^2 < K^2$, $\tilde{c}_{BG} > c_{BG}$, and the depth penetration length in the substrate is somewhat larger in the present case than in the electrically grounded solution of the preceding section. In many works, ε_0 is taken equal to one (thus granting same units to $\mathbf{D}$ and $\mathbf{E}$ (see the Lorentz–Heaviside unit system used in Maugin (1988)), and ε_{11} becomes the relative dielectric dielectric constant.

What about the associated quasi-particle? We have to evaluate (8.18) that we can rewrite as (*cf.* coloured Plate 8.2 with $U = A$)

$$\frac{d}{dt} \int_{V^+ + V^-} \mathbf{p}^w\, dV = \int_{V^+} div\, \mathbf{b}\, dV + \int_{V^-} div\, \mathbf{b}\, dV, \tag{8.72}$$

where integration among axis x_2 is now extended to the whole vertical band from minus to plus infinity (*cf.* coloured Plate 8.2).

As the mechanical part exists only in V^+, the evaluation of the left-hand side of this expression is formally just the same as in Paragraph B above. From the application of the divergence theorem (i) the terms at plus and minus infinity along x_2 will yield nothing because of the asymptotic conditions; (ii) the terms evaluated along the left L and right R sides, when combined will yield nothing as a consequence of periodicity, and (iii) the only terms to be reconsidered are those at the top interface between substrate and vacuum. The relevant term involves the quantity $\mathbf{n}^+ \cdot \mathbf{b}^+ + \mathbf{n}^- \cdot \mathbf{b}^- = \mathbf{n}^+ \cdot (\mathbf{b}^+ - \mathbf{b}^-) \equiv \mathbf{n}^+ \cdot [\mathbf{b}]$. Therefore, we have to pay attention to the components $-[b_{21}]$ and $-[b_{22}]$. Components b_{21} and b_{22} of tensor $\mathbf{b}$ are given by

$$b_{21} = -D_2\,\psi_{,1} = +\varepsilon_0\,\psi_{,2}\,\psi_{,1},$$

$$b_{22} = -L(x_2 = 0^-) + \varepsilon_0\,\psi_{,2}^2 = \frac{1}{2}\,\varepsilon_0\,\psi_{,2}^2, \qquad (8.73)$$

at $x_2 = 0^-$ because $L = \varepsilon_0\left(\psi_{,1}^2 + \psi_{,2}^2\right)/2$, and

$$b_{21} = +\varepsilon_{11}\,\psi_{,2}\,\psi_{,1}, \qquad b_{22} = -L(x_2 = 0^+) - D_2\,\psi_{,2}, \qquad (8.74)$$

at $x_2 = 0^+$. The second of these equations yields

$$b_{22} = -\frac{1}{2}\,\rho_0\left(\frac{\partial u_3}{\partial t}\right)^2 + \frac{1}{2}\,\bar{c}_{44}\left(u_{3,1}^2 + u_{3,2}^2\right) - \frac{1}{2}\,\varepsilon_{11}\left(\psi_{,1}^2 - \psi_{,2}^2\right). \qquad (8.75)$$

Computing the average over the x_1 interval, we note that the average of $-[b_{21}]$ will yield zero because only terms in $\sin\varphi \times \cos\varphi$ intervene. The average of $-[b_{22}]$ yields (compare to (8.54))

$$-\langle[b_{22}]\rangle(x_2 = 0) \propto \frac{1}{2}\,\rho_0\,A^2\,\widetilde{D}_{GB}(\omega, k_1) \equiv 0. \qquad (8.76)$$

As a final consequence, after collecting all results, the motion equation of the corresponding quasi-particle is given by (compare (8.61))

$$\frac{d}{dt}\left(\widetilde{M}_{BG}(A;\widetilde{K})\,\tilde{c}_{BG}\right) = 0, \qquad \widetilde{M}_{BG} = \frac{\rho_0\,\pi\,A^2}{2\,\widetilde{K}^2}. \qquad (8.77)$$

We do no repeat here the proof of the expression of the accompanying "kinetic energy" of the quasi-particle (but note that the integral of the energy must be taken over the whole vertical band including the energy of the free electric field in the region $x_2 < 0$). The resulting kinetic energy for the quasi-particle is Newtonian and given by

$$\widetilde{E}_{qp}(BG) = \frac{1}{2}\,\widetilde{M}_{BG}\,\tilde{c}_{BG}^2. \qquad (8.78)$$

As a partial conclusion, we recall that the case of BG surface waves (mechanically, pure SH waves, in spite of the presence of coupled electric effects) has been studied here because it nicely complements that of standard Rayleigh waves (mechanically, purely sagittaly polarized waves) of Chapter 7. Both correspond to *nondispersive* SAWs. In addition, this simple case allows us to study in the next sections the influence of possible elastic nonlinearity or viscosity in the substrate.

8.5   Perturbation by elastic nonlinearities

8.5.1   *Basic equations*

We look for the simplest nonlinear generalization of the model exploited in foregoing sections where the linear elasticity considered was equivalent to taking an elastic energy proportional to the quantity $\Phi = u_{3,1}^2 + u_{3,2}^2$. It happens that this is none other than a typical invariant in 2D isotropic elasticity (although we are working with an elastic crystal of 6mm symmetry with six-fold axis along the x_3-direction allowing for the existence of the piezoelectric coupling and, consequently, the existence of SH SAWs). Accordingly, the simplest generalization is to continue the expansion of the elastic energy in terms of Φ, e.g., to write an elastic energy of the form

$$W^{\text{elas}} = \frac{1}{2} c_{44} \, \Phi + \frac{1}{4} \beta \, c_{44} \, \Phi^2 + h.o.t. \tag{8.79}$$

Here c_{44} is the only remaining elasticity coefficient and β is a *nonlinearity parameter* that will be considered *small* in the sequel, higher-order terms (indicated by the ellipsis $h.o.t$) than the second order being discarded in (8.79). Higher-order nonlinear electroelastic couplings could also be introduced as done by Kalyanasundaram (1984) — also, Maugin *et al.* (1992b) — but this would bring nothing but useless complications to the modelling.

Compared to preceding sections, on account of (8.79) we now have to consider a bulk Lagrangian density of the form

$$L = \frac{1}{2} \rho_0 \, \dot{u}_3^2 + \frac{1}{2} \varepsilon_{11} \left(\psi_{,1}^2 + \psi_{,2}^2 \right) - \frac{1}{2} \bar{c}_{44} \left(\Phi + \frac{1}{2} \beta \, \Phi^2 \right), \tag{8.80}$$

and a bulk energy density

$$H = \frac{1}{2} \rho_0 \, \dot{u}_3^2 - \frac{1}{2} \varepsilon_{11} \left(\psi_{,1}^2 + \psi_{,2}^2 \right) + \frac{1}{2} \bar{c}_{44} \left(\Phi + \frac{1}{2} \beta \, \Phi^2 \right), \tag{8.81}$$

with

$$\bar{c}_{44} = c_{44} \, (1 + K^2), \qquad K^2 := \frac{e_{15}}{\varepsilon_{11} \, c_{44}}, \qquad \psi = \phi - \frac{e_{15}}{\varepsilon_{11}} u_3, \tag{8.82}$$

where ϕ is the original quasi-static electric potential and ψ is an effective scalar electric potential, and K is the electromechanical coupling coefficient.

It is readily shown than the remaining field equations are now replaced by the following system for $x_2 > 0$ [∇^2 is the two-dimensional Laplacian in the (x_1, x_2) plane]

$$\bar{c}_{44} \, \nabla^2 u_3 + \beta \, \bar{c}_{44} \left[(\Phi \, u_{3,1})_{,1} + (\Phi \, u_{3,2})_{,2} \right] = \rho_0 \, \ddot{u}_3, \qquad \nabla^2 \psi = 0. \tag{8.83}$$

The boundary conditions at $x_2 = 0$ are formally unchanged compared to the linear case (mechanically free surface, electroded surface put at zero potential ϕ), i.e.,

$$\mathbf{n} \cdot \boldsymbol{\sigma} = 0, \qquad \phi = \phi_0 = 0 \;\Rightarrow\; \psi = -\frac{e_{15}}{\varepsilon_{11}} u_3; \tag{8.84}$$

but, of course, the stress components have acquired nonlinear strain contributions. That is, now

$$\begin{aligned}
\sigma_{23} &= \bar{c}_{44} \left(1 + \beta\,\Phi\right) u_{3,2} + e_{15}\,\psi_{,2}, \\
\sigma_{13} &= \bar{c}_{44} \left(1 + \beta\,\Phi\right) u_{3,1} + e_{15}\,\psi_{,1},
\end{aligned} \tag{8.85}$$

and electric constitutive equations are left unchanged:

$$D_1 = -\varepsilon_{11}\,\psi_{,1}, \qquad D_2 = -\varepsilon_{11}\,\psi_{,2}, \qquad \phi = \psi + \frac{e_{15}}{\varepsilon_{11}}\,u_3. \tag{8.86}$$

The canonical conservation equations of wave momentum and energy are formally left unchanged. What is basically changed is the treatment of waves on account of the additional "quartic contribution" in the elastic energy because the body now becomes a generator of third harmonics.

8.5.2 *Surface wave solution*

We are in the situation described in the coloured Plate 8.3. The relevant equations to be solved are given by Equations (8.83) for the semi-infinite space $x_2 > 0$ and (8.84) at the upper surface $x_2 = 0$ with fields going to zero as $x_2 \to \infty$. Because of the considered type of nonlinearity and the smallness of the parameter β, we *a priori* consider solutions of the type:

$$\begin{aligned}
u_3 &= U \exp(-\alpha\,x_2) \cos(k\,x_1 - \omega\,t) \\
&\quad + \beta\,U^3 \exp(-3\,\alpha\,x_2) \cos\left(3\,(k\,x_1 - \omega\,t)\right) + O(\beta^2),
\end{aligned} \tag{8.87}$$

and

$$\begin{aligned}
\psi &= \Psi \exp(-\beta_\psi\,x_2) \cos(k\,x_1 - \omega\,t) \\
&\quad + \beta\,\Psi^3 \exp(-3\,\beta_\psi\,x_2) \cos\left(3\,(k\,x_1 - \omega\,t)\right) + O(\beta^2).
\end{aligned} \tag{8.88}$$

On substituting (8.87) into the first of (8.83) we obtain a bulk "dispersion relation" at order β as

$$\begin{aligned}
D(\omega, k, \alpha) &:= \omega^2 - \frac{\bar{c}_{44}}{\rho_0}\left[(k^2 - \alpha^2)\right. \\
&\quad \left. - \frac{\beta}{4} U^2 \exp(-2\,\alpha\,x_2)\,(9\,\alpha^4 - 3\,k^4 + 2\,k^2\alpha^2)\right] = 0,
\end{aligned} \tag{8.89}$$

for $x_2 > 0$. This still depends on x_2.

On substituting (8.88) into the second of (8.83) — which is **not** a propagation equation —, we obtain that

$$\beta_\psi = k, \tag{8.90}$$

so that (8.88) in fact reads

$$\begin{aligned}
\psi &= \Psi \exp(-k\,x_2)\,\cos(k\,x_1 - \omega\,t) \\
&\quad + \beta\,\Psi^3 \exp(-3\,k\,x_2)\,\cos\big(3\,(k\,x_1 - \omega\,t)\big) + O(\beta^2).
\end{aligned} \tag{8.91}$$

If $\beta = 0$, we recover the standard result for the Bleustein–Gulyaev wave.

We now apply the boundary conditions at $x_2 = 0$. From the second of these equations we obviously have:

$$\Psi = -\frac{e_{15}}{\varepsilon_{11}}\,U, \tag{8.92}$$

while from the other we arrive at

$$\left(\alpha - \overline{K}^2\,k\right) + \frac{\beta}{4}\,U^2\,\alpha\,(3\,\alpha^2 + k^2) = 0, \qquad \overline{K}^2 := \frac{K^2}{1 + K^2}. \tag{8.93}$$

In the linear case this would yield

$$\alpha = \alpha_0 = \overline{K}^2\,k_0, \tag{8.94}$$

corresponding to the dispersion relation (*cf.* Paragraph 8.4.2)

$$D_{BG}(\omega, k_0) := \omega^2 - c_{BG}^2\,k_0^2 = 0, \qquad c_{BG}^2 = \frac{\bar{c}_{44}^2}{\rho_0}\left(1 - \overline{K}^4\right). \tag{8.95}$$

In a weakly nonlinear case, we benefit from the smallness of β to envisage perturbed (k, α) solutions noted with a subscript S such that, at first order in β,

$$k_S = k_0 + \beta\,\bar{k}_S, \qquad \alpha_S = \alpha_0 + \beta\,\bar{\alpha}_S. \tag{8.96}$$

In view of (8.89) and (8.93), we need the second and fourth powers of these quantities. That is, at order one in β:

$$k_S^2 = k_0^2 + 2\,\beta\,k_0\,\bar{k}_S, \qquad \alpha_S^2 = \alpha_0^2 + 2\,\beta\,\alpha_0\,\bar{\alpha}_S. \tag{8.97}$$

Substituting from these into (8.89) and (8.93) at $x_2 = 0$, we obtain a system of two equations that we write in the following form:

$$\overline{K}^2\,\bar{k}_S - \bar{\alpha}_S = \overline{K}^2\,U^2\,k_0^3\left(1 - 3\,\overline{K}^4\right)/4 \equiv \tilde{s}_{01}, \tag{8.98}$$

$$2\left(\overline{K}^2\,\bar{\alpha}_S - \bar{k}_S\right) = U^2\,k_0^3\left(3 - 9\,\overline{K}^8 - 2\,\overline{K}^4\right)/4 \equiv \tilde{s}_{02}. \tag{8.99}$$

As $\overline{K}^4$ is small, both $\tilde{s}_{01}$ and $\tilde{s}_{02}$ are positive. The solution of this system is

$$\bar{k}_S = \frac{\overline{K}^2\,\tilde{s}_{01} + \tilde{s}_{02}/2}{\overline{K}^4 - 1} < 0, \qquad \bar{\alpha}_S = \frac{\tilde{s}_{01} + \overline{K}^2\,\tilde{s}_{02}/2}{\overline{K}^4 - 1} < 0. \tag{8.100}$$

In summary, together with (8.96), (8.87) and (8.91) we have thus obtained the nonlinear surface wave solution in the form

$$u_3 = U \exp(-\alpha_S x_2) \cos(k_S x_1 - \omega t)$$
$$+ \beta U^3 \exp(-3\alpha_S x_2) \cos\left(3\left(k_S x_1 - \omega t\right)\right) + O(\beta^2), \tag{8.101}$$

$$\psi = -\frac{e_{15}}{\varepsilon_{11}} U \exp(-k_S x_2) \cos(k x_1 - \omega t)$$
$$- \beta \frac{e_{15}^3}{\varepsilon_{11}^3} U^3 \exp(-3 k_S x_2) \cos\left(3\left(k_S x_1 - \omega t\right)\right) + O(\beta^2). \tag{8.102}$$

8.5.3 *Quasi-particle associated with the wave solution*

Following the strategy already applied in preceding paragraphs, we try to associate a so-called *quasi-particle* with the just obtained wave-motion by integrating the conservation laws over a material domain Ω that is representative of this wave motion. The so obtained *global* conservation equations read:

$$\frac{d}{dt} P_i = F_i, \quad i = 1, 2, \tag{8.103}$$

and

$$\frac{d}{dt} E = P_F, \tag{8.104}$$

wherein

$$P_i = \int_\Omega p_i^w \, d\Omega, \quad F_i := \int_{\partial\Omega} (\mathbf{n} \cdot \mathbf{b})_i \, dS, \tag{8.105}$$

and

$$E = \int_\Omega H \, d\Omega, \quad P_F := \int_{\partial\Omega} \mathbf{n} \cdot \left(\boldsymbol{\sigma} \cdot \dot{\mathbf{u}} + \mathbf{D}\,\dot{\phi}\right) dS. \tag{8.106}$$

The domain of integration Ω is selected as follows in Euclidean 3D space:

$$\Omega = [0, \lambda_S] \times [0, +\infty) \times [0, 1], \tag{8.107}$$

where $\lambda_S = 2\pi/k_S$ is the wavelength of the first harmonic component as altered by the nonlinearity. On account of (8.96) and the smallness of β, we have

$$\lambda_S = \frac{2\pi}{k_0}\left(1 - \beta \frac{\bar{k}_S}{k_0}\right) = \lambda_{BG}\left(1 - \beta \frac{\bar{k}_S}{k_0}\right), \tag{8.108}$$

where $k_0 = k_{BG}$ and λ_{BG} are the wave characteristics of the linear Bleustein–Gulyaev wave.

At the same order of approximation, we have the speed

$$c_S = \frac{\omega}{k_S} = c_{BG}\left(1 - \beta\,\frac{\bar{k}_S}{k_{BG}}\right),$$ (8.109)

while we note that

$$\frac{\bar{k}_S}{k_{BG}} < 0,$$ (8.110)

so that the corrections in (8.108) and (8.109) are *positive*.

On computing P_1 and P_2 from the first of definitions (8.105) and the solution (8.101)–(8.102), we obtain that

$$P_1 = \frac{\rho_0\,\pi\,\omega\,U^2}{2\,\overline{K}^2\,k_{BG}}\left(1 - \beta\,\frac{\bar{\alpha}_S}{\overline{K}^2\,k_{BG}}\right), \qquad P_2 \equiv 0,$$ (8.111)

where the second of these is a direct consequence of the periodicity of the first harmonic solution. The result $(8.111)_1$ can be rewritten in the following enlightening form:

$$P_1 = M_{BGNL}\,c_{BGNL},$$ (8.112)

where $c_{BGNL} = c_S$ and the "mass" M_{BGNL} is given by the expressions

$$M_{BGNL} = M_{BG}\left(1 - \frac{\beta}{4}\left(3\,\overline{K}^4 - 1\right)k_{BG}^2\,U^2\right)$$
$$= M_{BG}\left(1 - \beta\left(\frac{\bar{\alpha}_S - \overline{K}^2\,\bar{k}_S}{\overline{K}^2\,k_{BG}}\right)\right),$$ (8.113)

with M_{BG} the mass defined in (8.61). We note that the "speed" of the quasi-particle is increased as compared to the full linear case of Paragraph 8.4.3. Therefore, we would expect that the "mass" would be decreased. But, on examining the last part of (8.57) we observe that the two terms within parentheses perturb the usual mass at order β. The term proportional to $\bar{k}_S$ contributes to a decrease in the mass in agreement with an increase in velocity. However, the term involving $\bar{\alpha}_S$ has a counter effect because in the nonlinear case the attenuation in depth is less marked than in the linear case. Globally, these two antagonistic effects result in an increase of the mass.

Now we look for the possible right-hand side of the "quasi-particle" equation of motion (8.103), expecting it to vanish identically at our order of approximation. By definition

$$F_1 = \int_{\partial\Omega} (\mathbf{n}\cdot\mathbf{b})_1\,dS = -\int_0^{\lambda_S} b_{21}\Big|_{x_2=0}\,dx_1,$$
$$F_2 = -\int_0^{\lambda_S} b_{22}\Big|_{x_2=0}\,dx_1,$$ (8.114)

the other surface integrals providing nothing by virtue of the periodicity along x_1 and the bottom surface going to $+\infty$ with the corresponding fields tending towards zero. One easily shows that on account of the expression of L and of the relevant boundary condition we have

$$
\begin{aligned}
b_{21}\big|_{x_2=0} &= -D_2\,\phi_{,1}\big|_{x_2=0}, \\
b_{22}\big|_{x_2=0} &= -L(x_2=0) - D_2\,\phi_{,2}\big|_{x_2=0}.
\end{aligned}
\tag{8.115}
$$

From the first of these and the solution (8.102) we directly obtain that $F_1 \equiv 0$ on account of the periodicity of involved functions. As to the F_2 component, first we note that the dispersion relation (8.89) can be re-written at order β for $x_2 \geq 0$ as

$$
\begin{aligned}
D(\omega, k, \alpha) &= \omega^2 - \frac{\bar{c}_{44}}{\rho_0}\, k_{BG}^2 \left(1 - \overline{K}^4\right) \\
&\quad + 4\,\beta\, x_2\, \frac{\bar{c}_{44}}{\rho_0}\, k_{BG}^2\, \overline{K}^2 \left(\bar{k}_S - \overline{K}^2\, \bar{\alpha}_S\right) = 0,
\end{aligned}
\tag{8.116}
$$

an expression that obviously recurs to (8.95) in the absence of elastic non-linearity.

Then we evaluate the second of (8.115) on account of (8.88) and show that

$$
-\frac{1}{\pi} \int_0^{\lambda_{BG}} b_{22}\big|_{x_2=0}\, dx_1 \equiv 0,
\tag{8.117}
$$

on account of (8.116) applied at $x_2 = 0$ and the satisfaction of the dispersion relation at order zero in β.

As a result, both F_1 and F_2 are nil, and since $P_2 \equiv 0$, we are left with the unique inertial motion of the quasi-particle along the propagation direction of the wave, in the form

$$
\frac{d}{dt}\left(M_{BGNL}\, c_{BGNL}\right) = 0.
\tag{8.118}
$$

We must finally check, independently, that this is compatible with an equation of conservation of "kinetic energy" for the quasi-particle.

Conservation of energy

We apply equation (8.104) with the definitions (8.106) to the representative volume (8.107). The computation of the "power" P_F does not need to be reproduced as it is essentially the same as in the linear case (Paragraph 8.4.2) since only the stress is altered by the nonlinearity, while the boundary and limit condition are formally left unchanged. It is thus found that

$P_F \equiv 0$. Thus the energy E defined by $(8.106)_1$ will be conserved. We evaluate its expression at order β independently of the result (8.118) and will, afterwards, check that the two are compatible at that order. To this aim we substitute for the obtained surface wave solution in (8.105). The painstaking evaluation of this expression at order β leads to the following result:

$$
\begin{aligned}
E = {} & \frac{1}{2}\,\rho_0\,\omega^2\,U^2\,\frac{\pi}{2\,\alpha_S\,k_S} + \frac{1}{2}\,\bar{c}_{44}\left(\frac{k_S}{2\,\alpha_S} + \frac{\alpha_S}{2\,k_S}\right)U^2\,\pi \\
& + \frac{1}{4}\,\bar{c}_{44}\,\beta\,U^4\,\frac{(-1 + 4\,\overline{K}^4 - 3\,\overline{K}^8)}{8\,\overline{K}^2} - \overline{K}^2\,\bar{c}_{44}\,U^2\,\pi.
\end{aligned}
\tag{8.119}
$$

With some work this can be rewritten as

$$
E = \frac{1}{2}\,M_{BG}\,c_{BG}^2\left(1 + \beta\left(\frac{-2\,\bar{\alpha}_S}{\overline{K}^2\,k_{BG}} + \frac{U^2\,k_{BG}^2}{4}\,\frac{(-1 + 4\,\overline{K}^4 - 3\,\overline{K}^8)}{4\,(1 - \overline{K}^4)}\right)\right).
\tag{8.120}
$$

But we also note that at order β we can write

$$
\frac{1}{2}\,M_{BGNL}\,c_{BGNL}^2 = \frac{1}{2}\,M_{BG}\,c_{BG}^2\left(1 + \beta\left(\frac{\overline{M}_{BG}}{M_{BG}} + 2\,\frac{\bar{c}_{BG}}{c_{BG}}\right)\right),
\tag{8.121}
$$

with (*cf.* equations (8.113) and (8.109))

$$
\frac{\overline{M}_{BG}}{M_{BG}} = \frac{1}{4}\,(3\,\overline{K}^4 - 1)\,k_{BG}^2\,U^2 = \frac{\bar{\alpha}_S - \overline{K}^2\,\bar{k}_S}{\overline{K}^2\,k_{BG}},
\tag{8.122}
$$

$$
\frac{\bar{c}_{BG}}{c_{BG}} = -\frac{\bar{k}_S}{k_{BG}}.
\tag{8.123}
$$

Thus (8.121) reads

$$
\frac{1}{2}\,M_{BGNL}\,c_{BGNL}^2 = \frac{1}{2}\,M_{BG}\,c_{BG}^2\left(1 - \beta\left(\frac{\bar{\alpha}_S + \overline{K}^2\,\bar{k}_S}{\overline{K}^2\,k_{BG}}\right)\right),
\tag{8.124}
$$

or else

$$
\frac{1}{2}\,M_{BGNL}\,c_{BGNL}^2 = \frac{1}{2}\,M_{BG}\,c_{BG}^2\left(1 + \beta\left(\frac{-2\,\bar{\alpha}_S}{\overline{K}^2\,k_{BG}} + \frac{\bar{\alpha}_S - \overline{K}^2\,\bar{k}_S}{\overline{K}^2\,k_{BG}}\right)\right).
\tag{8.125}
$$

But on account of the first expression in (8.122) and of the identity $\big(-1 + 4\,\overline{K}^4 - 3\,\overline{K}^8\big) \equiv \big(3\,\overline{K}^4 - 1\big)\big(1 - \overline{K}^4\big)$, we find that (8.120) and (8.125) are exactly the same result. Accordingly, we can claim that, at order β, the kinetic energy of the quasi-particle computed from the general canonical definition (8.106) is not different from the one we can compute from the Newtonian mechanics of this quasi-particle once we know the expressions of the "mass" and of the velocity of this quasi-particle. It is readily checked that the kinetic energy of the quasi-particle in the nonlinear case is increased compared to the linear case.

In summary, we note that the "point mechanics" of the obtained quasi-particle is governed by the following equations of one-dimensional motion (in the propagation direction parallel to the limiting plane) and energy:

$$\frac{d}{dt}(M_{BGNL}\, c_{BGNL}) = 0, \qquad \frac{d}{dt}\left(\frac{1}{2}\, M_{BGNL}\, c_{BGNL}^2\right) = 0, \qquad (8.126)$$

that we proved exactly *independently* of one another at order β. Thus this mechanics is both Newtonian (first of (8.126)) and Leibnizian (second of (8.126)) with a particle kinetic energy that in fact results from all types (kinetic, elastic, electric and piezoelectric) of continuum energy densities. The obtained motion in (8.118) is inertial with no applied driving force and no import or expense of energy.

8.6 Perturbation by viscosity

8.6.1 *Some general words*

The study of the influence of the viscosity of the substrate on the propagation of standard Rayleigh waves has been a much debated problem to which a few publications refer (*cf.* Scholte (1947); Caloi (1950); Tsai and Kolsky (1968); Curie *et al.* (1977); Curie and O'Leary (1978); Romeo (2001a,b); Lai and Rix (2002); Carcione (2007)). Because of intrinsic difficulties in this problem, we here prefer to consider again the case of Bleustein–Gulyaev waves which, in spite of coupling with the electric potential, remains with only one elastic displacement polarized along thee x_3-axis. Papers considering this configuration with some dissipation are by Romeo (2001a,b) and Rousseau and Maugin (2011b). We base the following analysis on this last reference of which we recall the essential results in the next paragraph. Then the associated quasi-particle in non-inertial motion will be constructed thereafter. The considered situation will be the one illustrated in the coloured Plate 8.4.

8.6.2 *Reminder of the Bleustein–Gulyaev surface wave problem in presence of weak viscous losses*

For a semi-infinite space made of a linear piezoelectric and weakly viscous material with symmetry axis of order 6 orthogonal to the sagittal plane

(x_1, x_2), free of stress and electroded (zero potential) on its top limiting surface $x_2 = 0$, the approximate real SAW solution obtained in Rousseau and Maugin (2011b) is given for the SH (shear horizontal, orthogonal to the sagittal plane) elastic displacement u_3 and the effective electrostatic potential ψ by the rather general equations

$$u_3(x_1, x_2, t) = U \exp(-(k_{1d}^I x_1 + k_{2ud}^I x_2)) \cos(k_{1d}^R x_1 + k_{2ud}^R x_2 - \omega t),$$
$$(8.127)$$

$$\psi(x_1, x_2, t) = \Psi \exp(-(k_{1d}^I x_1 + k_{2\psi d}^I x_2)) \cos(k_{1d}^R x_1 + k_{2\psi d}^R x_2 - \omega t),$$
$$(8.128)$$

where superscripts I and R denote imaginary and real parts, respectively, and (U, Ψ) are the real amplitudes. Here, the employed constitutive equations read (electric relaxation is not envisaged for the sake of simplicity)

$$\sigma_{23} = \bar{c}_{44} u_{3,2} + \eta \dot{u}_{3,2} + e_{15} \psi_{,2},$$
$$\sigma_{13} = \bar{c}_{44} u_{3,1} + \eta \dot{u}_{3,1} + e_{15} \psi_{,1},$$
$$(8.129)$$

and

$$D_1 = e_{15} u_{3,1} - \varepsilon_{11} \phi_{,1}, \qquad D_2 = e_{15} u_{3,2} - \varepsilon_{11} \phi_{,2}, \qquad (8.130)$$

or

$$D_1 = -\varepsilon_{11} \psi_{,1}, \qquad D_2 = -\varepsilon_{11} \psi_{,2}, \qquad \phi = \psi + \frac{e_{15}}{\varepsilon_{11}} u_3. \qquad (8.131)$$

The only new quantities are the viscous stresses indicated by the viscosity η in (8.129). We remind the reader that a superimposed dot denotes the partial time derivative. An infinitesimally small parameter ε is introduced that compares the frequency of the propagating wave and the viscous frequency, and is such that

$$\varepsilon = \omega \tau, \qquad \tau = \frac{\eta}{\bar{c}_{44}}, \qquad \bar{c}_{44} = c_{44} + \frac{e_{15}^2}{\varepsilon_{11}}; \qquad (8.132)$$

It was shown in Rousseau and Maugin (2011b) that up to order ε the various wave numbers introduced in (8.127) and (8.128) are given by

$$k_{1d}^R = k_{BG}, \qquad k_{1d}^I = \varepsilon k_{BG} f_d = \varepsilon k_{BG1}, \qquad (8.133)$$

$$k_{2ud}^R = -\varepsilon \overline{K}^2 k_{BG} (1 + f_d) \equiv -\varepsilon \overline{K}^2 (k_{BG} + k_{BG1}),$$
$$k_{2ud}^I = \overline{K}^2 k_{BG}, \qquad (8.134)$$

$$k_{2\psi d}^R = -\varepsilon k_{BG} f_d \equiv -\varepsilon k_{BG1}, \qquad k_{2\psi d}^I = k_{BG}. \qquad (8.135)$$

The factor f_d present in these equations depends only on $\overline{K}$ and is given by

$$f_d = \frac{1 + \overline{K}^4}{2 (1 - \overline{K}^2)}. \qquad (8.136)$$

The wave number k_{BG} relates to the corresponding nondissipative case recalled in Paragraph 8.4.2. So in principle we know this approximate solution which, we emphasize, is *not purely periodic* in propagation space. Therefore, there will be a cumulative effect of attenuation as the wave propagates. We cannot expect the associated quasi-particle to be in inertial motion.

8.6.3 Global equations of wave momentum and energy

In the present case (viscous losses alone), the local ("non")-conservation laws of energy and wave momentum (8.21) and (8.22) are reduced to

$$\frac{\partial}{\partial t}\left(\frac{1}{2}\rho_0\,\dot{\mathbf{u}}^2 + W\right) - \frac{\partial}{\partial x_j}\left(\sigma_{ji}\,\dot{u}_i + D_j\,\dot{\phi}\right) = -\sigma_{ji}^{\text{visco}}\,\dot{u}_{i,j}, \qquad (8.137)$$

and

$$\frac{\partial}{\partial t}p_i^w - \frac{\partial}{\partial x_j}b_{ji} = -\sigma_{jk}^{\text{visco}}\,u_{k,ji}, \qquad (8.138)$$

where the following quantities have been defined:

$$W = \overline{W}(e_{ji} = u_{(j,i)},\ E_j = -\phi_{,j}), \qquad (8.139)$$

$$\sigma_{ji} = \frac{\partial \overline{W}}{\partial e_{ji}} + \sigma_{ji}^{\text{visco}}, \quad D_j = -\frac{\partial \overline{W}}{\partial E_j}, \qquad (8.140)$$

$$p_i^w := -\rho_0\,\dot{u}_j\,u_{j,i}, \quad b_{ji} = -\left(L\,\delta_{ji} + \sigma_{jk}\,u_{k,i} + D_j\,\phi_{,i}\right),$$
$$L = \frac{1}{2}\rho_0\,\dot{\mathbf{u}}^2 - W. \qquad (8.141)$$

For the specially considered electroelastic motion (u_3, ϕ or ψ), function of (x_1, x_2, t) only, and $x_2 > 0$, we have the specific expressions (ρ_0 is a constant matter density)

$$L = \frac{1}{2}\rho_0\,\dot{u}_3^2 + \frac{1}{2}\varepsilon_{11}\left(\psi_{,1}^2 + \psi_{,2}^2\right) - \frac{1}{2}\bar{c}_{44}\left(u_{3,2}^2 + u_{3,1}^2\right), \qquad (8.142)$$

and

$$H = \frac{1}{2}\rho_0\,\dot{u}_3^2 - \frac{1}{2}\varepsilon_{11}\left(\psi_{,1}^2 + \psi_{,2}^2\right) + \frac{1}{2}\bar{c}_{44}\left(u_{3,2}^2 + u_{3,1}^2\right), \qquad (8.143)$$

for the Lagrangian and Hamiltonian energy densities per unit material volume, respectively.

The relevant boundary conditions (vanishing normal stress, imposed zero electric potential) at $x_2 = 0$ are the same as before.

$$\mathbf{n}\cdot\boldsymbol{\sigma} = 0, \quad \phi = 0 \ \Rightarrow\ \Psi + \frac{e_{15}}{\varepsilon_{11}}U = 0. \qquad (8.144)$$

Like in other sections, we consider first Equation (8.138) and look for its volume integral over an appropriately chosen material volume Ω and boundary $\partial\Omega$ equipped with unit outward normal $\mathbf{n}$ i.e., on account of the divergence theorem,

$$\frac{d}{dt}\int_\Omega p_i^w \, d\Omega - \int_{\partial\Omega} n_j \, b_{ji} \, dA = \int_\Omega F_i \, d\Omega, \tag{8.145}$$

where F_i is a source "material" force here due to viscosity:

$$F_i = -\sigma_{jk}^{\text{visco}} \, u_{j,ki} = -\left(\eta \, \dot{u}_{3,1} \, u_{3,1i} + \eta \, \dot{u}_{3,2} \, u_{3,2i}\right), \quad i = 1, 2, \tag{8.146}$$

while, just the same as before:

$$p_i^w = -\rho_0 \, \dot{u}_k \, u_{k,i} = \left(p_1^w = -\rho_0 \, \dot{u}_3 \, u_{3,1}, \; p_2^w = -\rho_0 \, \dot{u}_3 \, u_{3,2}\right). \tag{8.147}$$

For the global energy equation, from (8.137), we will have to evaluate the various terms in the balance equation

$$\frac{d}{dt}\int_\Omega H \, d\Omega - \int_{\partial\Omega} \mathbf{n} \cdot \left(\boldsymbol{\sigma} \cdot \dot{\mathbf{u}} + \mathbf{D} \, \dot{\phi}\right) dA = -\int_\Omega \sigma_{ji}^{\text{visco}} \, \dot{u}_{i,j} \, d\Omega. \tag{8.148}$$

Now, the crux of the matter is the selection of the volume Ω supposed to be representative of the wave motion. As noticed before, the obtained SAW solution is *not purely periodic* in propagation space. This requires serious thinking about the selection of the volume Ω. Its thickness along x_3 will be unity. Along the depth direction x_2 it will naturally extend from the free surface $x_2 = 0$ to plus infinity, with all fields tending towards zero in this limit after the very nature of a surface wave. As to the propagation direction x_1, while in the nondissipative case, any interval of one wavelength along the path could define the integration interval, in the present case we propose the following strategy. Let λ_{1d} the wavelength of the periodic part. Then the left side face L_n (see coloured Plate 8.4) of the path will be taken at an abscissa equal to n times this wavelength and the right face R_n of the path at $n + 1$ times the wavelength. Summing over a number of such intervals will account for the attenuation effect. Accordingly, the elementary volume of integration Ω is selected as

$$\Omega = [n \, \lambda_{1d}, (n + 1) \, \lambda_{1d}] \times [0, +\infty) \times [0, 1]. \tag{8.149}$$

Note that this will be a true volume integral with application of the divergence theorem — but there is no flux through the two surfaces parallel to the sagittal plane. We shall first proceed to the evaluation of (8.145) at the order ε. Detailed computations are given in Rousseau and Maugin (2012a). Only essential results are recalled, referring the interested reader to the detailed computations given in that paper.

A.   Evaluation of the x_1 component (usual propagation space)

First, retaining only terms of order ε at most, after some lengthy algebra we obtain that

$$P_{QPL} := \int_\Omega p_1^w \, d\Omega = M_{BGd} \, c_{BG}, \qquad (8.150)$$

wherein $c_{BG} = \omega/k_{BG}$ is the speed of the nondissipative solution, and we have defined a time-dependent "mass" (per unit thickness in the x_3 direction; detail of the proof in Appendix 2 of Rousseau and Maugin (2012a)) by

$$
\begin{aligned}
M_{BGd} &= M_{BGd}\big(\overline{K}, n; t\big) \\
&:= M_{BG} \exp\left\{ -4\,\pi\,\varepsilon\,f_d\left(n + \frac{1}{2}\right) \right\} \left[1 + \varepsilon\, f_d \sin(2\,\omega\,t)\right],
\end{aligned}
\qquad (8.151)
$$

where $M_{BG} = \rho_0 \, \pi \, U^2 / 2\,\overline{K}^2$ is the "mass" of the quasi-particle associated with the nondissipative Bleustein–Gulyaev SAW (*cf.* equation (8.61)).

Second, we evaluate the right-hand side of equation (8.145) in the x_1 direction. It is found that

$$
\begin{aligned}
B_1 &:= \int_\Omega F_1 \, d\Omega = \frac{\varepsilon\,\rho_0\,\omega^2\,U^2\,\pi\,f_d}{\overline{K}^2\,k_{BG}} \exp\big(-2\,\alpha(n)\big) + O(\varepsilon^2) \\
&= 2\,\varepsilon\,\omega\,P_{BG}\,f_d \exp\big(-2\,\alpha(n)\big),
\end{aligned}
\qquad (8.152)
$$

where we have set

$$\alpha(n) = 2\,\pi\,\varepsilon\,f_d\left(n + \frac{1}{2}\right), \qquad P_{BG} = M_{BG}\,c_{BG}, \qquad (8.153)$$

the last quantity being the momentum of the quasi-particle in the nonviscous case.

Finally, we must compute the projection onto the direction x_1 of the second term in the left-hand side of (8.145). Since there is no flux through the two faces of Ω parallel to the sagittal plane, and the quantities at $x_2 = +\infty$ all tend toward zero, we are left with the flux terms on the top surface S_1 and the fluxes through the side faces L_n and R_n in Plate 8.4. A tedious computation yields for

$$\overline{A}_1 := \int_{R_n} b_{11}\big((n+1)\,\lambda_{1d}\big) \, dx_2 - \int_{L_n} b_{11}(\lambda_{1n}) \, dx_2, \qquad (8.154)$$

the estimate (*cf.* Appendix 2 in Rousseau and Maugin (2012a))

$$
\begin{aligned}
\overline{A}_1 &= -\frac{\varepsilon\,\rho_0\,\omega^2\,U^2\,\pi\,f_d}{\overline{K}^2\,k_{BG}} \big(1 - \cos(2\,\omega\,t)\big) \exp\big(-2\,\alpha(n)\big) + O(\varepsilon^2) \\
&= -2\,\varepsilon\,\omega\,P_{BG}\,f_d \big(1 - \cos(2\,\omega\,t)\big).
\end{aligned}
\qquad (8.155)
$$

On the other hand, it can be shown that

$$\widetilde{A}_1 = - \int_{n\,\lambda_{1d}}^{(n+1)\,\lambda_{1d}} b_{21}\big|_{x_2=0} \, dx_1 \equiv 0, \tag{8.156}$$

where

$$b_{21}\big|_{x_2=0} = -D_2 \, \phi_{,1}\big|_{x_2=0}. \tag{8.157}$$

As a result, by computing the time derivative of P_{QPL} we *exactly* check that at order ε there holds the *identity*

$$\frac{d}{dt} P_{QPL} \equiv A_1 + B_1 =: F_L, \qquad A_1 = \overline{A}_1 + \widetilde{A}_1 \equiv \overline{A}_1, \tag{8.158}$$

with,

$$F_L = 2\,\varepsilon\,\omega\, P_{BG}\, f_d \, \exp\big(-2\,\alpha(n)\big)\,\cos(2\,\omega\,t), \tag{8.159}$$

where the two sides of $(8.158)_1$ were calculated independently of one another.

Remark 8.1. The factor $n+1/2$ present in the expressions (8.151), (8.153), (8.154), and (8.159) is tantamount to suggesting that within the interval of integration along x_1, the associated quasi-particle is "physically" located in the middle of this interval, midway between L_n and R_n.

B.   *Evaluation of the x_2 component (depth space)*

In the nondissipative case (Paragraph 8.4.2 above) it was shown that the x_2 component of equation (8.145) — with vanishing right-hand side — yields an identity $0 = 0$. The situation is markedly different in the present dissipative case because there remains to evaluate the following quantities. First,

$$P_{QPT} := \int_\Omega p_2^w \, dV = - \int_\Omega \rho_0 \, \dot{u}_3 \, u_{3,2} \, d\Omega, \tag{8.160}$$

with the evaluation

$$\begin{aligned} p_2^w = \rho_0\,\omega\,U^2\, &\exp\big(-2\,\varepsilon\,k_{BG1}\,x_1\big)\,\exp\big(-2\,\overline{K}^2\,k_{BG}\,x_2\big) \\ &\big(\overline{K}^2\,k_{BG}\,\sin X\,\cos X - \varepsilon\,(k_{BG}+k_{BG1})\,\sin^2 X\big), \end{aligned} \tag{8.161}$$

where

$$X = k_{BG}\,x_1 - \varepsilon\,\overline{K}^2\,(k_{BG}+k_{BG1})\,x_2 - \omega\,t. \tag{8.162}$$

Second, we have

$$\widetilde{A}_2 := - \int_{S_1} b_{22}(x_2 = 0)\, dx_1, \tag{8.163}$$

where, as easily checked, on account of the homogeneous boundary condition for $\boldsymbol{\sigma}$ at $x_2 = 0$,

$$b_{22}\big|_{x_2=0} = -\left(L - \varepsilon_{11}\,\psi_{,2}^2 - e_{15}\,\psi_{,2}\,u_{3,2}\right)\big|_{x_2=0}. \tag{8.164}$$

Note that there is no accompanying source quantity B_2 such as

$$B_2 = -\eta \int_\Omega \left(\dot{u}_{3,1}\,u_{3,12} + \dot{u}_{3,2}\,u_{3,22}\right)\,d\Omega, \tag{8.165}$$

because the integrand involving terms in $\sin(2X)$, hence terms that are $O(\varepsilon)$, will finally yield a contribution B_2 of order ε^2 compared to the other terms of order ε in the x_2 component of (8.146). Now we evaluate $\widetilde{A}_2$ given by (8.163) and find that (*cf.* Appendix 2 in Rousseau and Maugin (2012a))

$$\widetilde{A}_2 = -\frac{\varepsilon\,\rho_0\,\omega^2\,U^2\,\pi\,f_d}{k_{BG}}\,\exp\left(-\,2\,\alpha(n)\right)\,\sin(2\,\omega\,t). \tag{8.166}$$

Finally, for the x_2 (transverse, in depth) component of the quasi-particle motion, we obtain an equation with all terms of order ε as

$$\frac{d}{dt}P_{QPT} = A_2 =: F_T, \qquad A_2 \equiv \widetilde{A}_2, \tag{8.167}$$

with

$$P_{QPT} = \varepsilon\,P_{BG}\,\overline{K}^2\,m(t), \tag{8.168}$$

$$m(t) := -1 + f_d\left[\cos(2\,\omega\,t) - 1\right], \tag{8.169}$$

and (at order ε)

$$F_T = -2\,\varepsilon\,\omega\,P_{BG}\,\overline{K}^2\,f_d\,\exp\left(-\,2\,\alpha(n)\right)\,\sin(2\,\omega\,t). \tag{8.170}$$

We check that the latter is nothing but the time derivative of P_{QPT}, which indeed is of order ε.

C. Summary

Summing up, Equations (8.158) and (8.167) are the two components of the motion of the considered quasi-particle. Clearly, at order ε, the motion of this quasi-particle has become *two-dimensional* in the sagittal place. But we must remark that, on account of the smallness of some parameters, P_{QPL} is always positive for a right-traveling wave, and its time derivative, or F_L, is always of the sign of $\cos(2\,\omega\,t)$ for $t \in (0, 2\,\pi/2\,\omega)$. In comparison, P_{QPT} is always negative and its time derivative, hence F_T, is always of the sign of $(-\sin(2\,\omega\,t))$ for $t \in (0, 2\,\pi/2\,\omega)$. Furthermore, on comparing the expressions of F_L and F_T of the "*driving forces*" acting on the quasi-particle we obtain the following identity:

$$\overline{K}^4 F_L^2 + F_T^2 = \left(2\,\varepsilon\,\omega\,P_{BG}\,\overline{K}^2\,f_d\right)^2 \exp\left(-4\,\alpha(n)\right). \tag{8.171}$$

The right-hand side of this expression is a constant (of order ε^2) multiplied by the factor $\exp(-4\,\alpha(n))$. Thus, the relation (8.171) can be granted the following interpretation: the two "force" components (F_L, F_T) are such that the end point of the formed vector in the sagittal plane Π_S describes an ellipse — in the counter clockwise direction — of decreasing size (with n) as time grows in the course of propagation of the associated quasi-particle along x_1.

D.   *Global equation of energy for the quasi-particle*

We have to integrate equation (8.137) over the appropriate material volume, in principle the representative volume defined in (8.149) for the viscous case. We shall do this at the order ε at most. But first we remark the following: at this order the phase velocity in the x_1 direction is the same as in the nondissipative case, being equal to c_{BG}. Furthermore, the problem in the x_2 direction is of higher order than in the x_1 direction so that this in depth behavior (component P_{QPT} of the quasi particle momentum) can be discarded. The longitudinal velocity of the quasi-particle, just like the phase velocity of the continuum solution equal to c_{BG}, is a constant. Thus the energy equation for the "surface" quasi-particle can simply be obtained (same reasoning as for the theorem of kinetic energy in classical mechanics) by multiplying both sides if Equation (8.158) by c_{BG}. The latter being constant, we trivially obtain the energy equation as

$$\frac{d}{dt}(P_{QPL}\,c_{BG}) \equiv \frac{d}{dt}(M_{BGd}\,c_{BG}^2) = A_1\,c_{BG} + B_1\,c_{BG}. \tag{8.172}$$

This result is *not standard* because of the absence of factor $1/2$ in the left-hand side and the fact that the "mass" varies in time (due to viscous effects) while the velocity is constant. We shall return to this point here below because we must also show that Equation (8.172) can also be obtained independently of (8.158) by really integrating (8.137) over the volume (8.149). This direct evaluation is reported in Rousseau and Maugin (2012a). It indeed yields (8.172). This energy equation is unusual due to the constancy of the velocity, because it in fact reads

$$\left(\frac{dM_{BGd}}{dt}\right)c_{BG}^2 = \left(A_1 + B_1\right)c_{BG}. \tag{8.173}$$

Obviously, a nonvanishing B_1 is the direct result of the presence of viscous dissipation. But it is also true that a nonzero A_1 is also due to viscous dissipation. In effect, in the absence of dissipation, both the mass and the velocity of the quasi-particle become constant (inertial motion of a particle of invariant mass). Then the energy theorem takes the classical "kinetic energy" form with factor $1/2$. In the case of (8.173) the power of "applied forces" (here due to viscosity) results in the variation of Leibniz's (of 1686) "vis viva" (twice the standard kinetic energy). The fact that mass varies with time in that case correlates with the classical view that the amplitude of the signal decreases with propagation in the viscous medium in the continuous description (the "mass" of the quasi-particle is proportional to the square of this amplitude) while we know that the phase velocity is affected at a higher order only (*cf.* Rousseau and Maugin (2011b)). An analysis at this higher order would be of interest but we do not envisage it here.

One final remark is that despite the "Leibnizian" outlook of Equation (8.172), the equation of motion still looks "Newtonian" but with a varying mass in time. There are other cases where such a fact happens, e.g., in relativistic physics where mass increases with velocity, becoming infinite as the particle velocity tends towards the light velocity (so-called "Lorentzian" behavior), and also models of nonlinear dispersive waves where the wave-momentum equation yielding a quasi-particle representation also involves a varying mass, sometimes in an "anti-Lorentzian" manner (mass decreasing with velocity — see Maugin and Christov (2002, pp. 142–143), for such pathological cases). In the present case, mass decreases because of the influence of dissipation, while velocity is kept constant at the retained order of approximation.

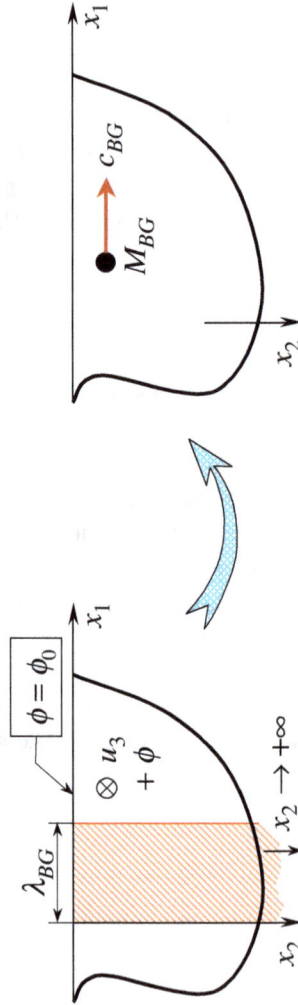

Plate 8.1   The standard Bleustein-Gulyaev SAW and its associated quasi-particle.

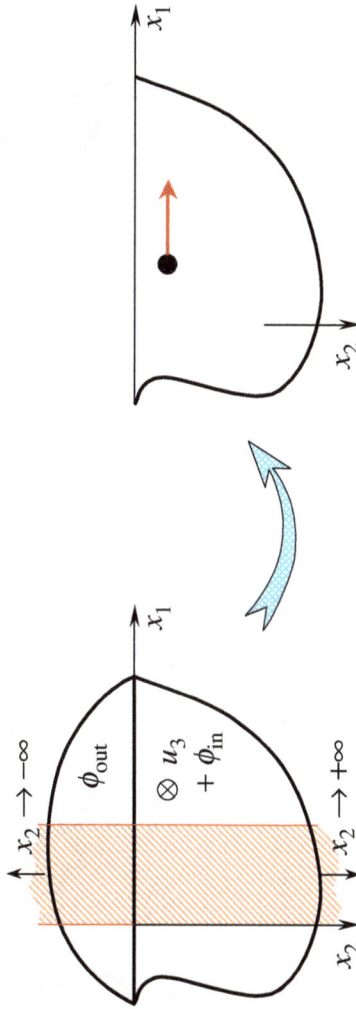

Plate 8.2 Bleustein–Gulyaev SAW: matching with an external electric field.

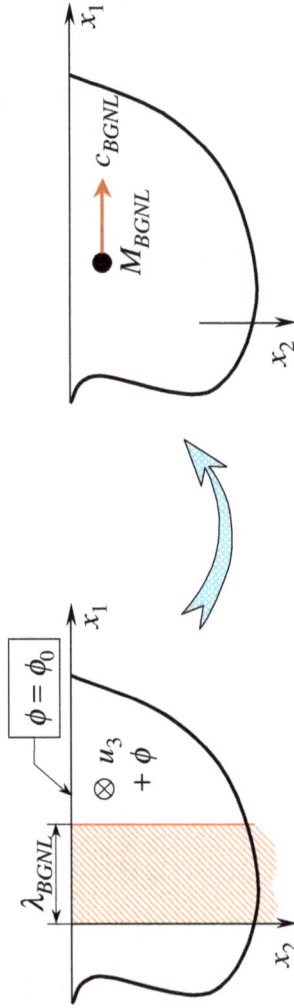

B-G SAW's on weakly nonlinear elastic substrate

Nonlinearity of the quartic type with coefficient β

$$\lambda_{BGNL} = \lambda_{BG}\left(1 - \beta\,\overline{k}_s/k_{BG}\right)$$

$\phi = \phi_0$

$\otimes u_3$

$+ \phi$

λ_{BGNL}

x_1

x_2

x_1

c_{BGNL}

M_{BGNL}

x_2

$$\frac{\mathrm{d}}{\mathrm{d}t}\left(M_{BGNL}\,c_{BGNL}\right) = 0, \quad M_{BGNL} = M_{BG}\left(1 - \beta\left(\frac{\overline{\alpha}_s - \overline{K}^2\,\overline{k}_s}{\overline{K}^2\,k_{BG}}\right)\right), \quad \frac{\mathrm{d}}{\mathrm{d}t}\left(\frac{1}{2}M_{BGNL}\,c_{BGNL}^2\right) = 0$$

Plate 8.3   Bleustein–Gulyaev SAW: influence of the substrate nonlinearity.

B-G SAW's on weakly viscous elastic substrate

→ Low losses of the Newtonian fluid type with viscosity η

$$\varepsilon_\nu = \omega\,\tau, \qquad \tau = \eta/\bar{c}_{44}, \qquad \bar{c}_{44} = c_{44}\,(1+\bar{K}^2)$$

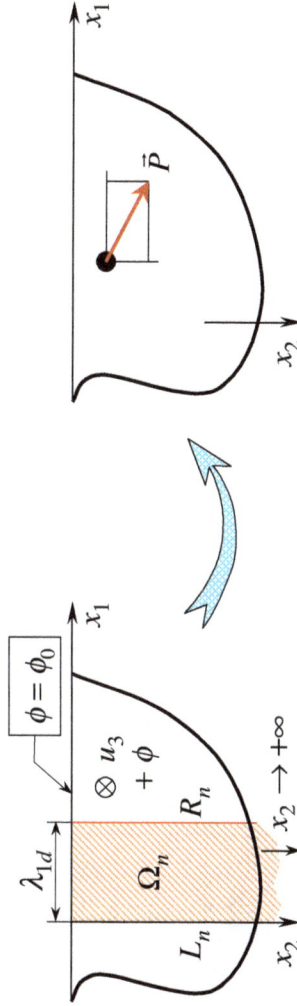

$$\frac{\mathrm{d}}{\mathrm{d}t}P_{QPL} = F_L, \quad P_{QPL} = M_{BG}\,g(\bar{K},n;t),$$

$$\frac{\mathrm{d}}{\mathrm{d}t}P_{QPT} = F_T, \quad P_{QPT} = \varepsilon_\nu\,P_{BG}\,\bar{K}^2\,m(t), \quad F_T = \mathrm{O}(\varepsilon_\nu)$$

$$\left(\frac{\mathrm{d}}{\mathrm{d}t}M_{BGd}\right)c_{BG}^2 \neq 0$$

Plate 8.4   Bleustein–Gulyaev SAW: Influence of the substrate viscosity.

Chapter 9

Waves in generalized elastic continua

Object of the Chapter

In this chapter, the possibility of associating the notion of quasi-particles with elastic wave modes is explored for three basic models of generalized continua: strain-gradient model (weak nonlocality), elasticity with a microstructure such as in Cosserat/micropolar materials, and a true nonlocal model (strong nonlocality) involving the long-range interactions in the underlying crystal lattice. In each case a simplified one-dimensional model is considered. Approximate solutions involving small parameters and exhibiting scale effects are obtained for the Newtonian-like motion of associated quasi-particles. Interpretation for alternate wave-like and quasi-particle-like behaviours is given.

9.1 The notion of generalized continuum

In this chapter we are interested in generalized elastic continua. The notion of generalized deformable continuum has been delineated in previous works (e.g., Maugin (2013, Chap. 13)). The chosen characterization of such continua is that their kinematic and kinetic description goes farther than the standard one introduced in his foundational work by Cauchy in 1822 (*cf.* Maugin (2011a); Maugin (2014, Chap. 3)). That is, in describing such elastic media we shall generally need more than the notions of displacement gradient, strain, symmetric Cauchy stress, and forces. We may also say beforehand that the additionally introduced descriptors correspond to a more inclusive look at a material body at a smaller scale than usual allowing for an evidence of a microstructure and the existence of internal length scales (the latter do not exist in standard linear elasticity). The last mentioned point is of course of importance when studying wave processes that are themselves characterized by a length scale (the wavelength), whence a possible competition or resonance, or else while studying elastic field singularities (at cracks, dislocations, *etc.*) characterized by an influence length (*cf.* Lazar (2012); Lazar and Maugin (2005, 2007)). Three possible avenues have been identified in a strictly deterministic approach.

The first generalization is that which describes the local spatial displacement variation with a higher degree of accuracy than usual in the standard theory of elasticity. This is referred to as *gradient theories*, the second gradient of the displacement or first gradient of the strain being generally sufficient to arrive at some new effective improvement. This clearly is of interest wherever the strain is not spatially uniform. Such considerations go back to A. Barré de Saint-Venant (1869) and J. Le Roux (1911, 1913) — and perhaps Piola earlier in the nineteenth century — and in more recent times to R. D. Mindlin and H. F. Tiersten (1962) and R. A. Toupin (1964). This is often referred to as a *weakly nonlocal approach*.

The second path consists in the introduction of a microstructure at each material point, in the simplest case, a rigidly rotating one just like in so-called *Cosserat continua*. This idea indeed goes back to the Cosserat brothers (1909), with more modern presentations and full dynamics by, e.g., A. C. Eringen (1968) and W. Nowacki (1986), under the name of *micropolar continua*. Deformable microstructure has been envisaged by R. D. Mindlin (1964) and Suhubi and Eringen (1964), and is often referred to to the mechanics of *micromorphic continua*.

The third path is called that of *strongly nonlocal* theories of continua where the mechanical response at a material point depends on a larger spatial domain than the immediate neighbourhood of the considered point. The idea may be traced back all the way to P. Duhem (1893). Syntheses on this type of approach are given by A. I. Kunin (1982) and A. C. Eringen (2002) although other scientists involved in its development include D. Rogula, D. G. B. Edelen, E. Kröner, *etc.* This is strongly correlated with a view to accounting for typical crystal properties (e.g., dispersion) while remaining in a continuum framework.

In what follows we shall examine how the notion of wave momentum applies to these different schemes after having recalled the basic equations for each case and giving a short reminder of wave propagation in such media. The exposed results are due to the authors (Rousseau and Maugin (2013)).

9.2 Weak nonlocality and gradient model of elasticity

9.2.1 *Summary of strain-gradient elasticity*

We use the Cartesian tensor notation and Einstein summation convention over repeated indices. According to Mindlin's theory of strain-gradient elasticity the local equation of linear momentum reads — just the same as in the standard theory — in the absence of body force:

$$\rho_0 \frac{\partial^2 u_i}{\partial t^2} - \frac{\partial}{\partial x_j} \sigma_{ji} = 0, \tag{9.1}$$

where the Cauchy stress is given by

$$\sigma_{ji} = \bar{\sigma}_{ji} - \frac{\partial}{\partial x_k} m_{kji} \equiv \frac{\delta W}{\delta e_{ij}}, \tag{9.2}$$

with

$$\bar{\sigma}_{ji} = \frac{\partial W}{\partial e_{ij}}, \qquad m_{kji} = \frac{\partial W}{\partial (e_{ji,k})}, \qquad W = W(e_{ij}, e_{ij,k}), \tag{9.3}$$

and $\delta/\delta e_{ij}$ denotes a functional derivative. A very particular form of W in the *linear* case is proposed as (*cf.* Aifantis (1992))

$$W = \frac{1}{2} \bar{\sigma}_{ji} \, e_{ij} + \frac{1}{2} \delta^2 \frac{\partial}{\partial x_k} (\bar{\sigma}_{ji}) \frac{\partial}{\partial x_k} (e_{ji}), \tag{9.4}$$

where δ is a characteristic length. This means that once $\bar{\sigma}_{ji}$ is formally known, so is the case of m_{kji}, and then only δ is a new material parameter compared to the standard linear elasticity. The working hypothesis (9.4) is used so as to obtain the simplest generalization of standard elasticity. This is sufficient to study the characteristic singularities (dislocations, disclinations) appearing in elasticity problems. Fortunately, we do not need here the associated boundary conditions that may be quite farfetched (*cf.* Germain (1973)).

In the present theory, the *conservation of wave momentum* in a homogeneous body (no body forces) in general reads (for a homogeneous material; *cf.* Maugin and Trimarco (1992))

$$\frac{\partial}{\partial t} p_i^w - \frac{\partial}{\partial x_j} b_{ji} = 0, \tag{9.5}$$

with wave momentum p_i^w and Eshelby material tensor b_{ji} given by (*cf.* Maugin and Trimarco (1992); Kalpakides and Agiasofitou (2002))

$$p_i^w = -\rho_0 \, \dot{u}_j \, u_{j,i}, \tag{9.6}$$

$$b_{ji} = -(L \, \delta_{ji} + \bar{\sigma}_{jp} \, u_{p,i} + 2 \, m_{jpq} \, u_{q,pi}) + (m_{jpq} \, u_{p,i})_{,q}, \tag{9.7}$$
$$L \equiv K - W, \qquad K = \rho_0 \, \dot{\mathbf{u}}^2/2.$$

This follows from a direct application of Noether's theorem. An interesting reference on symmetries and conservation laws in the present theory is Lazar and Anastassiadis (2007).

9.2.2   *Wave solution for a simplified problem*

We consider the relevant equations recalled in the foregoing paragraph. However, we further simplify the problem by considering a propagation along axis $x_1 = x$, no dependence on the orthogonal coordinate x_2, and the only involved elastic displacement is a transverse one, $u_3 = u$ (had we considered a surface wave, u_3 would be a shear-horizontal (SH) component). With the above simplifications, Equation (9.2) yields the only nonvanishing stress component

$$\sigma = \sigma_{31} = \mu \, u_{,x} + \alpha \, \mu \, \delta^2 \, u_{,xxx}, \tag{9.8}$$

where we set $c_{44} = \mu$ and introduced a numerical factor $\alpha = \pm 1$ which allows us to play with the sign of a possible dispersion. This matter is thoroughly discussed by Askes and Aifantis (2011).

Equation of motion (9.1) yields the dispersion relation

$$\omega^2 = c_0^2 \, k^2 \, (1 + \alpha \, \delta^2 \, k^2), \qquad c_0^2 = \frac{\mu}{\rho_0}, \qquad k_0 \equiv \frac{\omega}{c_0}. \tag{9.9}$$

If $\delta = 0$, we are in standard linear elasticity for which we introduce the notation $k_0 = \omega/c_0$ and $\lambda_0 = 2\pi/k_0$. The case $\delta \neq 0$ but obviously small or relatively small compared to a characteristic macroscopic length, requires a short discussion about the possible dispersion. Let λ the wave length of the solution. We can envisage two cases: (i) a resonance between the scales of λ and δ, i.e., $\lambda \approx \delta$ with $\lambda \ll \lambda_0$, or (ii) $\delta \ll \lambda \ll \lambda_0$. The case $\lambda < \delta$ is not meaningful. We note $c = \omega/k$ the phase velocity of the solution. With the above convention for α, we have the following possibilities:

$$c \equiv c_0 \left(1 \pm \frac{\delta^2 \, k^2}{2} \right), \quad k \equiv k_0 \left(1 \mp \frac{\delta^2 \, k^2}{2} \right), \quad \lambda \cong \lambda_0 \left(1 \pm \frac{\delta^2 \, k^2}{2} \right). \tag{9.10}$$

The possibility with the upper signs in these expressions disagrees with the working hypotheses because we must have $\lambda \ll \lambda_0$ and also $c < c_0$. We also remark that $|\lambda - \lambda_0| = O(\delta^2 \, k^2) \, \lambda_0 \ll 1$. The case $\lambda \approx \delta$ is thus excluded and we are left with the case $\delta \ll \lambda \ll \lambda_0$. Note that the choice of the present signs means that in the energy density in (9.4) the plus sign in the second contribution must be replaced by a minus sign. This brief discussion agrees with the usual occurrence of anomalous dispersion as obtained in the Boussinesq equation of crystal physics (*cf.* Maugin (1999b, p. 46)).

9.2.3   *Wave momentum*

On account of the above choice, we now compute the sum over a representative volume element of the wave process (one wave length in the propagation

direction; a square of unit sides in the transverse plane) of the surviving
component of the wave momentum (9.6). We obtain thus

$$P^w = \rho_0 \, \pi \, \omega \, U^2, \qquad (9.11)$$

where U is the amplitude. This can be written in the following alternate
forms of a momentum of a quasi-particle:

$$P^w = M_0 \, c_0 = M \, c, \qquad (9.12)$$

where

$$M_0 = \rho_0 \, \pi \, k_0 \, U^2, \qquad (9.13)$$

$$M = M_0 \left(1 + \frac{1}{2} \delta^2 \, k_0^2 \right), \quad c = c_0 \left(1 - \frac{1}{2} \delta^2 \, k_0^2 \right). \qquad (9.14)$$

Accordingly, $M > M_0$. This means that in the second expression in
the right-hand side of (9.12) the quantity of information contained in the
"mass" M is larger. But the two possibilities in (9.12) are possible only
because the difference between the two expressions vanishes at our order of
approximation (as immediately checked by computing the product of the
two quantities in (9.14)). The difference between the quasi-particles associated with the standard (nondispersive) elastic case and the approximate
gradient case is $O\big((\delta^2 \, k^2)^2\big)$ and escapes the present approximation. This
case, therefore, is moderately interesting. Of greater interest is the case of
Cosserat continua.

9.3 The case of Cosserat continua

9.3.1 *Summary of the general linear theory*

This theory can be extracted from Eringen (1968) or Maugin (2011b, Chap.
9), in the following form (no externally applied body force and couple).
Local balance of linear momentum:

$$\rho_0 \, \frac{\partial^2 u_i}{\partial t^2} - \frac{\partial}{\partial x_j} \sigma_{ji} = 0, \qquad (9.15)$$

Local balance of angular momentum:

$$\rho_0 \, j_{ij} \, \frac{\partial^2 \phi_j}{\partial t^2} - \frac{\partial m_{ji}}{\partial x_j} - \varepsilon_{ipq} \, \sigma_{pq} = 0, \qquad (9.16)$$

where ϕ_j are the vector components of a small rotation angle, m_{ji} are the
components of the couple-stress, and ε_{ipq} is the alternating symbol. Obviously, the Cauchy stress tensor of components σ_{ji} is no longer symmetric.

For the sake of simplicity the micro-inertia j is taken isotropic, $j_{ji} = I\,\delta_{ij}$. The elastic constitutive equations are given by

$$\boldsymbol{\sigma} = \frac{\partial \widehat{W}}{\partial \mathbf{e}}, \qquad \mathbf{m} = \frac{\partial \widehat{W}}{\partial \boldsymbol{\gamma}}, \tag{9.17}$$

with strain measures $\mathbf{e}$ and $\boldsymbol{\gamma}$

$$\begin{aligned}
\mathbf{e} &:= (\nabla\mathbf{u})^T + \text{dual } \phi = \{e_{ji} = u_{i,j} - \varepsilon_{jk}\,\phi_k\}, \\
\boldsymbol{\gamma} &:= \nabla\phi = \{\gamma_{ji} = \phi_{i,j}\}.
\end{aligned} \tag{9.18}$$

On account of the above introduced approximations, the canonical balance equation of momentum takes on the following form for a homogeneous material (Maugin (1998)) where the nonlinear theory is given for finite strains and finite micro-rotations)

$$\frac{\partial}{\partial t} p_i^w - \frac{\partial}{\partial x_j} b_{ji} = 0, \tag{9.19}$$

wherein

$$p_i^w = -\rho_0 \left(\dot{u}_j\, u_{j,i} + I\,\dot{\phi}_j\, \phi_{j,i} \right), \tag{9.20}$$

and

$$b_{ji} = -\left(L\,\delta_{ji} + \sigma_{jk}\, u_{k,i} + m_{jk}\, \phi_{k,i} \right), \tag{9.21}$$

with a Lagrangian density now given by

$$L = \frac{1}{2}\, \rho_0\, \dot{\mathbf{u}}^2 + \frac{1}{2}\, \rho_0\, I\, \dot{\phi}_k\, \dot{\phi}_k - \widehat{W}. \tag{9.22}$$

Note that although (9.19) basically is an equation of linear momentum (resulting from the invariance of the *whole* physical system under consideration by translations of the coordinate parametrization in space (i.e., material space in modern jargon), both the wave momentum (9.20) and the Eshelby stress (9.21) involve the internal degree of freedom of rotation simultaneously with the usual displacement. Equations (9.19) through (9.21) follow directly from the canonical expression of Noether's theorem applied to the basic fields (u_i, ϕ_i) with a visible summation over terms generated by these two fields.

For linear isotropic homogeneous Cosserat continua, equations (9.15) and (9.16) read (*cf.* Eringen (1999, p. 124))

$$(\lambda + 2\mu + \kappa)\,\nabla(\nabla\cdot\mathbf{u}) - (\mu + \kappa)\,\nabla\times\nabla\times\mathbf{u} + \kappa\,\nabla\times\phi = \rho_0\,\ddot{\mathbf{u}}, \tag{9.23}$$

and

$$(\alpha + \beta + \gamma)\,\nabla(\nabla\cdot\phi) - \gamma\,\nabla\times\nabla\times\phi + \kappa\,(\nabla\times\mathbf{u} - 2\,\phi) = \rho_0\,I\,\ddot{\phi}, \tag{9.24}$$

where λ and μ are usual Lamé coefficients, coefficient κ clearly accounts for the coupling between displacement and micro-rotation, and coefficients α, β, and γ relate to micro-rotation effects. The non-negativeness of the energy potential imposes the following sets of inequalities:

$$3\lambda + 2\mu + \kappa \geq 0, \qquad 2\mu + \kappa \geq 0, \qquad \kappa \geq 0, \tag{9.25}$$

and

$$3\alpha + \beta + \gamma \geq 0, \qquad \gamma + \beta \geq 0, \qquad \gamma - \beta \geq 0. \tag{9.26}$$

9.3.2 *Wave solution for a simplified problem*

For the sake of illustration, we consider a very special case of the above-recalled formulation. The propagation takes place along the $x_1 = x$ axis. The unique elastic displacement $u_3 = u$ is transverse. It couples with the micro-rotation whose axis is along the remaining transverse direction $x_2 = y$. The remaining two coupled field equations issued from (9.23) and (9.24) read

$$(\mu + \kappa) \nabla \times \nabla \times \mathbf{u} - \kappa \nabla \times \boldsymbol{\phi} + \rho_0 \, \ddot{\mathbf{u}} = 0, \tag{9.27}$$

$$\gamma \nabla \times \nabla \times \boldsymbol{\phi} - \kappa \left(\nabla \times \mathbf{u} - 2 \, \phi \right) - \rho_0 \, I \, \ddot{\boldsymbol{\phi}} = \mathbf{0}, \tag{9.28}$$

with $\mathbf{u} = u \, \mathbf{e}_z$ and $\boldsymbol{\phi}$ reduced a single component ϕ along $\mathbf{e}_y$, or

$$c_2^2 \frac{\partial^2}{\partial x^2} u + \frac{1}{2} I \omega_0^2 \frac{\partial \phi}{\partial x} - \frac{\partial^2}{\partial t^2} u = 0, \tag{9.29}$$

$$c_4^4 \frac{\partial^2}{\partial x^2} \phi - \omega_0^2 \, \phi + \frac{1}{2} \omega_0^2 \frac{\partial u}{\partial x} - \frac{\partial^2}{\partial t^2} \phi = 0, \tag{9.30}$$

where we have set

$$c_2^2 = \frac{\mu + \kappa}{\rho_0}, \qquad c_4^2 = \frac{\gamma}{\rho_0 \, I}, \qquad \omega_0^2 = \frac{2 \, \kappa}{\rho_0 \, I}, \qquad c_T^2 = \frac{\mu}{\rho_0}. \tag{9.31}$$

Wave solution

For wavelike solutions of amplitudes (U, Φ) for (u, ϕ), frequency ω, and wave number k, equations (9.29) and (9.30) provide the system

$$\left(\omega^2 - c_2^2 \, k^2 \right) U - \frac{i}{2} I \omega_0^2 \, k \, \Phi = 0, \tag{9.32}$$

$$\frac{i}{2} \omega_0^2 \, k \, U + \left(\omega^2 - \omega_0^2 - c_4^2 \, k^2 \right) \Phi = 0. \tag{9.33}$$

The dispersion relation provided by these two equations is given by

$$D(\omega, k) = 4 \left(\omega^2 - c_2^2 \, k^2 \right) \left(\omega^2 - \omega_0^2 - c_4^2 \, k^2 \right) - I \omega_0^4 \, k^2 = 0. \tag{9.34}$$

Amplitudes (U, Φ) are related by the equation

$$\Phi = -\frac{2\,i}{I\,\omega_0^2\,k}\,(\omega^2 - c_2^2\,k^2)\,U. \tag{9.35}$$

The first term in the left-hand side of (9.32) shows that the essentially transverse displacement mode deviates from the classical nondispersive shear elastic mode given by $\omega^2 - c_T^2\,k^2 = 0$. The second term in the left-hand side of (9.33) shows that the essentially micro-rotation mode will be such that $\omega^2 \geq \omega_0^2$. This is a high-frequency mode similar to an optical mode in crystals. The surviving transverse acoustic (TA) branch has a slope equal to c_T at the origin of the dispersion diagram and a slope tending to c_2 for large k's. The group velocity varies non-monotonically between these two values — see Figure 5.11.2 in Eringen (1999, p. 150).

9.3.3   *Wave momentum and quasi-particles*

In the present case, the wave momentum (9.20) has a nonvanishing component only along the propagation direction, i.e.,

$$p^w = -\rho_0\,(u_{,t}\,u_{,x} + I\,\phi_{,t}\,\phi_{,x}). \tag{9.36}$$

Furthermore, we consider in this expression only the *acoustic (TA) branch*. It is rendered dispersive because of its coupling with micro-rotation, and this is true only for small or relatively small wave numbers (long wave lengths). We now compute the sum over a representative volume element of the wave process (one wave length in the propagation direction; a square of unit sides in the transverse plane) of the wave momentum (9.36) on account of the relation (9.35). This yields a momentum per unit volume in the form

$$P^w = \rho_0\,\pi\,\omega\,|U|^2\left(1 + 4\,\frac{(\omega^2 - c_2^2\,k^2)}{I\,\omega_0^4\,k^2}\right). \tag{9.37}$$

This can be rewritten in the form of the product of a frequency-dependent "mass" and a velocity, as for a *quasi-particle*, in the canonical form

$$P^w = M_T(\omega)\,c, \tag{9.38}$$

where we have set

$$M_T(\omega) = \rho_0\,\pi\,k\,|U|^2\left(1 + 4\,\frac{(\omega^2 - c_2^2\,k^2)}{I\,\omega_0^4\,k^2}\right), \tag{9.39}$$

or, equivalently,

$$M_T(\omega) = \rho_0\,\pi\,k\,|U|^2\left(1 + \frac{\omega^2 - c_2^2\,k^2}{\omega^2 - \omega_0^2 - c_4^2\,k^2}\right). \tag{9.40}$$

The two crucial parameters here are the micropolar modulus κ and the micro-inertia I.

Relations (9.38) through (9.40) can be re-interpreted thus. Since a Cosserat medium may be thought of as a kind of granular material with a characteristic size d of granules, we can envisage two cases: (i) $\lambda \gg d$, and we can set $I = 0$ without much loss, in which case we have the following reduction (standard elasticity but with shear coefficient from the above equations):

$$M_{T_0}(\omega) = \rho_0\, \pi\, |U|^2\, \frac{\omega}{c_2} < M_{T_0} = \rho_0\, \pi\, |U|^2\, \frac{\omega}{c_T}, \qquad (9.41)$$

with $c_T < c_2$ since $\kappa > 0$. We are in some kind of standard elasticity with a slighty more rigid body.

The second case (ii) is $\lambda \cong d$, in which case we have resonance with the microstructure (the cases $\lambda \ll d$ and $\omega \to +\infty$ are not meaningful). But as $\omega \to 0$, we find that

$$M_T(\omega) \to M_{T_0}(\omega) \left(1 + \frac{I\,\omega^2}{4\,c_T^2}\right), \qquad (9.42)$$

so that $M_T(\omega) > M_{T_0}(\omega)$. This means that we have more information contained in one wavelength than for the case without microstructure (a greater mass contains more information than a smaller one in our general vision of the "mass" of quasi-particles). If we re-write (9.39) symbolically as

$$M_T(\omega) = \rho_0\, \pi\, |U|^2\, k(\omega)\, (1 + A(\omega)), \qquad (9.43)$$

we have

$$M_{T_0}(\omega) = \rho_0\, \pi\, |U|^2\, k_T\, (1 + A_{\max}) < M_T(\omega). \qquad (9.44)$$

This is the "quasi-particle" interpretation of the TA mode in Cosserat continua for small wave numbers.

9.4 The case of strong nonlocality

9.4.1 *Summary of nonlocal elasticity*

We consider only small strains. The balance of linear momentum still has the form (9.1). But a typical linear elastic constitutive equation is given by an expression of the type

$$\sigma_{ji}(\mathbf{x}, t) = C_{jikl}^0\, e_{kl}(\mathbf{x}, t) + \int_D C_{jikl}^*(\mathbf{x}, \mathbf{x}')\, e_{kl}(\mathbf{x}', t)\, d\mathbf{x}', \qquad (9.45)$$

where C^0_{jikl} is the tensor of local elasticity coefficients while C^*_{jikl} is that of nonlocal elasticity coefficients, and D is a spatial domain to which both points $\mathbf{x}$ and $\mathbf{x}'$ belong. Equation (9.45) can also be written as

$$\sigma_{ji}(\mathbf{x}, t) = \int_D \left[C^0_{jikl}\, \delta(\mathbf{x} - \mathbf{x}') + C^*_{jikl}(\mathbf{x}, \mathbf{x}') \right] e_{kl}(\mathbf{x}', t)\, d\mathbf{x}', \qquad (9.46)$$

where δ stand for Dirac's generalized function. As a special case of this we can envisage the following expression

$$\sigma_{ji}(\mathbf{x}, t) = \int_D \left[C^0_{jikl}\, \alpha(|\mathbf{x} - \mathbf{x}'|) \right] e_{kl}(\mathbf{x}', t)\, d\mathbf{x}', \qquad (9.47)$$

where α is a function of influence (or kernel of nonlocality) that decreases sufficiently fast and D is a domain of influence (in principle, the whole space) around point $\mathbf{x}$. In one space dimension, (9.47) reduces to

$$\sigma(x, t) = \int_{-\infty}^{+\infty} E\, \alpha(|x - x'|)\, \frac{\partial u(x', t)}{\partial x'}\, dx', \qquad (9.48)$$

where E is a unique elasticity coefficient. The datum of the function α specifies the type of nonlocality. When α, that satisfies translational invariance, is none other than the delta function we recover classical (local) elasticity.

In general (9.47) can be deduced from a potential *via* a functional (space) derivative (see Eringen (2002)). This allows one to build the balance of wave momentum by application of Noether's theorem to the corresponding variational formulation. The nonlocality does not formally affect the expression of the wave momentum which, per unit volume, still reads

$$p^w_i(\mathbf{x}, t) = -\rho_0\, \dot{u}_j(\mathbf{x}, t)\, u_{j,i}(\mathbf{x}, t). \qquad (9.49)$$

9.4.2   *Wave solution for a simplified problem*

For the sake of illustration we consider the very special case given by a nonlocal constitutive equation of the type (9.48). The remaining scalar equation of motion reads (*cf.* Eringen, 1972, 1974, 1987)

$$\rho_0 \frac{\partial^2 u}{\partial t^2} - \frac{\partial}{\partial x} \left[\int_{-\infty}^{+\infty} E\, \alpha(|x - x'|)\, \frac{\partial u(x', t)}{\partial x'}\, dx' \right] = 0, \qquad (9.50)$$

or

$$\frac{\partial^2 u}{\partial t^2} - c_0^2 \frac{\partial}{\partial x} \left[\alpha * \frac{\partial u}{\partial x} \right] = 0, \qquad (9.51)$$

where the symbol $*$ stands for the convolution product (in space) and we have set $c_0^2 = E/\rho_0$.

Wave solution

We look for progressive wave solutions of (9.51) and introduce for that purpose Fourier transforms in space and time

$$\bar{u}(k,t) = \frac{1}{2\,\pi} \int_{-\infty}^{+\infty} u(x,t) \, \exp(-i\,k\,x) \, dx, \tag{9.52}$$

$$\tilde{u}(x,\omega) = \frac{1}{2\,\pi} \int_{-\infty}^{+\infty} u(x,t) \, \exp(i\,\omega\,t) \, dt, \tag{9.53}$$

so that

$$\tilde{\bar{u}}(k,\omega) = \frac{1}{(2\,\pi)^2} \iint u(x,t) \, \exp\left(-i\,(k\,x - \omega\,t)\right) dx\,dt. \tag{9.54}$$

On applying this to (9.51) we find that

$$\left(\omega^2 - \omega_0^2 \, \bar{\alpha}(k)\right) \tilde{\bar{u}} = 0, \qquad \omega_0^2 := c_0^2 \, k^2. \tag{9.55}$$

Now the important choice is that of α or $\bar{\alpha}$. Inspired by lattice theory and following other authors (e.g., Eringen (2002)) we suggest, as a worthy example, to take (square of the cardinal sine)

$$\bar{\alpha}(k) = \frac{4}{k^2\,a^2} \, \sin^2\left(\frac{k\,a}{2}\right), \tag{9.56}$$

so that, trivially, by taking the inverse transform,

$$\alpha(x) = 1 - \left|\frac{x}{a}\right|, \qquad -a \leq x \leq +a. \tag{9.57}$$

The latter is a triangle or "hat" function, and a can be identified as a lattice parameter, while the interval in the second of (9.57) looks like a Brillouin zone. For a nonvanishing solution, (9.55) now delivers the dispersion relation in the form

$$\omega^2(k) = \frac{4\,c_0^2}{a^2} \, \sin^2\left(\frac{k\,a}{2}\right). \tag{9.58}$$

This is indeed the dispersion relation in the Born–Kármán (1954) one-dimensional model of a lattice with nearest neighbour interactions with proper identification of $E = K/a$ and $m = \rho_0\,a^3$ if m is the individual mass of the "atoms" in the linear chain and K is the spring constant between neighbours. Here the nonlocality is manifested by the dispersion corresponding to the simplest discrete model of elasticity. We can already surmise that the associated wave momentum will be peculiar.

9.4.3 *Wave momentum and quasi-particles*

For a one-dimensional motion the local wave momentum takes the standard form

$$p^w = -\rho_0 \, u_{,t} \, u_{,x}. \tag{9.59}$$

We could pursue the above applied strategy by taking the double transform (9.54) of this quantity, yielding

$$\tilde{\tilde{p}}^w = -\rho_0 \, \omega \, k \, \tilde{\tilde{u}} * \tilde{\tilde{u}}, \tag{9.60}$$

and thus, by extracting $\omega \, k$ from (9.58), we would have

$$\tilde{\tilde{p}}^w(k,\omega) = -\rho_0 \, c_0 \, \frac{\sin(k \, a/2)}{k \, a/2} \, (k^2 \, \tilde{\tilde{u}} * \tilde{\tilde{u}}). \tag{9.61}$$

But then we have to return to physical space where the solution will involve Heaviside functions as a result of expressions (9.57). It is simpler to consider directly (9.50) with a real periodic wave mode $u = U \cos(k \, x - \omega_0 \, t)$, so that taking account of (9.58) and evaluating the average $\langle p^w \rangle_\lambda$ of the result over a representative volume element of length one wave length in the propagation direction and cross section of unit sides, we obtain the global wave momentum in the form

$$P^w := \langle p^w \rangle_\lambda = \rho_0 \, \omega_0 \, \pi \, U^2 \, \frac{\sin(k \, a/2)}{k \, a/2} = \rho_0 \, k \, c_0 \, \pi \, U^2 \, \frac{\sin(k \, a/2)}{k \, a/2}. \tag{9.62}$$

This can be rewritten in the following two alternate forms:

$$P^w := \langle p^w \rangle_\lambda = M \, c_0 = M_0 \, c, \tag{9.63}$$

where we have set

$$M_0 = \rho_0 \, k \, \pi \, U^2, \tag{9.64}$$

and

$$M = M_0 \, \frac{\sin(k \, a/2)}{k \, a/2} < M_0, \qquad c = c_0 \, \frac{\sin(k \, a/2)}{k \, a/2} < c_0, \tag{9.65}$$

where c_0 is the elastic wave speed of the nondispersive case (classical linear elasticity). Equation (9.64) is none other than the formula for the mass of the associated quasi-particle for this classical linear elasticity (as previously used in the transmission-reflection problem; *cf.* Chapter 5 above).

In the limit of large wavelength ($\lambda \gg \pi \, a$, i.e., $(k \, a/2) \ll 1$) we recover the standard linear elasticity case as

$$P_0^w = M_0 \, c_0. \tag{9.66}$$

But the dispersive case resulting in (9.63) admits two different interpretations. In terms of the wave-propagation picture the nonlocality affects the velocity via the cardinal sine function that creates dispersion. But in the quasi-particle picture it is the quantity of information, hence the mass, that is influenced by the nonlocality. The interest must be focused on the first of the two representations (9.63).

In the discussion of the dispersive case we should account for the fact that there exists a minimal wavelength that corresponds to accord between the wavelength and the lattice spacing, i.e., $\lambda = 2a < \pi a$. Below this resonance condition the description does not make sense. Accordingly, the velocity c and the mass M remain in the following intervals:

$$c_{\min} \leq c \leq c_0, \qquad c_{\min} = \frac{2}{\pi} c_0, \tag{9.67}$$

and

$$M_{\min} \leq M \leq M_0, \qquad M_{\min} = 2\rho_0 \pi \frac{U^2}{a}. \tag{9.68}$$

As $\lambda < \lambda_0$ in the dispersive case, the integration domain in propagation space contains a smaller number of particle points in the lattice. As a consequence, less information pertaining to the lattice vibration is contained in M than in M_0, hence a justification for the validity of the first of (9.68) for the mass of the quasi-particle. It might be possible to relate these considerations to those of "phonons" (viz. the interpretation of "pseudo-momentum" — our wave momentum — and material velocity given by Nelson (1979, p. 56), and the considerations about "quasi-momentum" by Gurevich and Thellung (1990)). What we note, however, is that (9.62) can also be written as

$$P^w := \langle p^w \rangle_\lambda = A k, \qquad A \equiv \rho_0 c_0 \pi U^2 \frac{\sin(k a/2)}{k a/2} = \rho_0 c \pi U^2, \tag{9.69}$$

where A may be called the elementary *wave action* (energy multiplied by time). It is of interest to compare the first of (9.69) to the expression of the value of the total energy obtained in the same conditions, that is, by comparing its value integrated over the same representative volume element. The elementary kinetic energy for the wave solution is given by

$$K = \frac{1}{2} \rho_0 \left(\frac{\partial u}{\partial t} \right)^2 = \frac{1}{2} \rho_0 \omega^2 U^2 \sin^2(k x - \omega t), \tag{9.70}$$

This yields

$$K^w := \langle K \rangle_\lambda = \frac{1}{2} A \omega. \tag{9.71}$$

For the model (9.50) of nonlocal elasticity, the elementary strain (or potential) energy is given by

$$W = \frac{1}{2} \int_D E\,\alpha(|x - x'|)\,\frac{\partial u(x, t)}{\partial x}\,\frac{\partial u(x', t)}{\partial x'}\,dx',\qquad(9.72)$$

where D is the domain of influence, here given by the interval $[-a, +a]$, and α is given by (9.57). The evaluation of the value of the integral of (9.72) over the representative volume element here is given within square brackets).

[Introducing nondimensional space variables by $X = x/a$, $X'' = X - X'$, for the wave solution $u = U\,\cos(k\,x - \omega\,t)$ we can write (9.70) in the form

$$W = \frac{1}{2}\,E\,k^2\,U^2\,I\,\sin(k\,x - \omega\,t),\qquad(a)$$

with

$$I := \int_{-1}^{+1} (1 - |X''|)\,\sin(k\,x - \omega\,t - k\,a\,X'')\,dX''.\qquad(b)$$

This is nothing but

$$\begin{aligned}
I = {} & \int_{-1}^{+1} \sin(k\,x - \omega\,t - k\,a\,X'')\,dX'' \\
& + \int_{-1}^{0} (+X'')\,\sin(k\,x - \omega\,t - k\,a\,X'')\,dX'' \\
& + \int_{0}^{1} (-X'')\,\sin(k\,x - \omega\,t - k\,a\,X'')\,dX''.
\end{aligned}$$

The careful computation of these three integrals leads to the following result:

$$I = \sin(k\,x - \omega\,t)\,\frac{\sin^2(k\,a/2)}{(k\,a/2)^2}.\qquad(c)$$

Substituting from this into (a) and taking the volume integral over the wave representative volume element of length $\lambda = 2\,a$ in the propagation direction and square transverse cross section of sides of unit length, we obtain

$$\langle W \rangle_\lambda = \frac{1}{2}\,\rho_0\,\pi\,U^2\,c^2\,k = \frac{1}{2}\,\rho_0\,\pi\,c\,U^2\,\omega,\qquad(d)$$

where c is given by (9.65)$_2$.]

Remarkably enough, we obtain

$$W^w := \langle W \rangle_\lambda = \frac{1}{2}\,A\,\omega.\qquad(9.73)$$

Thus we have *equipartition of energy* or, in other words, a *vanishing* wave Lagrangian $L^w = K^w - W^w$. From this there follows that the total wave energy is given by

$$E^w = K^w + W^w = A\omega. \qquad (9.74)$$

On comparing the two results (9.70) and (9.74) we see that there holds the relation

$$\widetilde{A}^w := P^w x - E^w t = A\varphi, \qquad (9.75)$$

where $\varphi = kx - \omega t$ is the phase, and $\widetilde{A}^w$ may be called the quasi-particle *wave action*. This last result is formally identical to the combination of Planck's and de Broglie's elementary relations, $E = \hbar\omega$ and $p = \hbar k$ in the form $px - Et = \hbar\varphi$, where $\hbar$ is Planck's reduced constant or quantum of action (see Chapter 4). Thus our dispersive wave system in the model of nonlocal elasticity is interpreted in the framework of quasi-particles just as for the case the elementary particles of quantum-wave mechanics physics. A discussion of such dual properties was already emphasized in Maugin and Rousseau (2010a) for the Lighthill–Whitham wave mechanics where the vanishing of the wave Lagrangian results from the satisfaction of the dispersion relation.

9.5 Conclusion

In the foregoing three sections we have answered the program proposed at the outset, sometimes only at some degree of approximation allowed by the presence of small parameters. The three studied cases of course allow for the possible introduction of a Newtonian mechanics of quasi-particles (QP) associated in the three models of generalized continua, that is: (i) a simple model of strain-gradient elasticity (so-called weak nonlocality) *à la* "*Aifantis*", (ii) a one-dimensional (but two-degrees of freedom) model of Cosserat/microplar continuum — where only the transverse acoustic mode is envisaged for the QP association, but altered by its coupling with a high frequency rotational mode akin to an "optical" mode — , and (iii) a one-dimensional model of (strongly) nonlocal elasticity that initially reproduces the Born–Kármán lattice with dispersion. In the last case the interactions between the mode wavelength and the lattice structure play a fundamental role in the interpretation of the mass and velocity of the associated quasi-particle, in particular when there is resonance between the two scales (wavelength and lattice structure).

But the question of the consideration of surface waves (i.e., waves with amplitude and energy essentially localized in the vicinity of a guiding flat surface) — which provided the impetus for our original studies (e.g., Maugin and Rousseau (2012a) — was left unanswered. As we know we can proceed in this direction only if we know before hand the analytical wave-like solutions. These solutions are *not* simple, but attempts at these have been given for the three cases of generalized continua considered here, respectively in works by Georgiadis *et al.* (2004) and Vardoulakis and Georgiadis (1997), Suhubi and Eringen (1964), and Eringen (1973). The corresponding association of quasi-particles guided by the limiting surface of the half space remains to be studied.

Chapter 10

Examples of solitonic systems

Object of the Chapter

This last chapter complements the preceding ones in giving examples of systems involving both dispersion and nonlinearity simultaneously, and thus prone to accept the propagation of more or less true solitons. Here equations of associated quasi-particles are obtained by integration of the conservation laws over the whole real line in the direction of propagation since the notion of wavelength has disappeared. Systems now standard in soliton theory are recalled in the framework of crystal elasticity. These are 1D spatial systems of varied complexity. A more complex surface acoustic wave problem is evoked by way of conclusion, exhibiting thus the power of some of the asymptotic methods and wave-kinematic considerations given in Chapter 4.

10.1 Introduction: The notion of soliton

In the last four decades, solitons have become a true paradigm of nonlinear physics, and the theory of solitons has become a field of expertise in its own right (*cf.* Newell (1985); Drazin and Johnson (1989); Infeld and Rowlands (1990); Dauxois and Peyrard (2006)). The name *soliton* itself, coined by Martin Kruskal and his co-workers in the 1960s, reflects the mathematical property that true (mathematically defined) such strongly localized nonlinear waves do behave like massive point particles in interactions with fellow waves (elastic collision, no change in dynamic properties except for a change of phase). This is mathematically transcribed in the property of exact integrability of such systems — and the appropriately constructed theory of inverse scattering by Kruskal *et al.* Many physicists are less demanding; they are satisfied with the obvious properties of individual solitary waves which they often call solitons by abuse of language. While the mathematical properties of strictly defined solitons result in the existence of an infinite hierarchy of conservation laws associated with the initial model equation, the physicist's definition is related to the existence of a few conservation (or *non*-conservation) equations. This fits well in the view expanded in the present volume. It also permits to deal elegantly with non-exactly integrable systems — obviously more frequent in physical reality than mathematically integrable systems — as examined in the critical synthesis of Kivshar and Malomed (1989). In the present chapter, we adopt this viewpoint as it agrees with our view of wave momentum and the associated equations of motion of quasi-particles and the accompanying energy equations. To achieve such an agenda we first recall the standard cases of solitons, and then apply the formalism to more original cases that generalize some models already considered in Chapters 7 and 8 and exploit some of the techniques introduced in preceding chapters (e.g., Chapters 3 and 4).

10.2   Reminder: Some standard cases

10.2.1   *The Boussinesq model in elastic crystals*

An often cited exemplary equation in the present context is the so-called *Boussinesq equation* which has both fluid-mechanics and crystal-mechanics origins (*cf.* Christov *et al.* (2007)). In appropriate units, this may be considered as the field equation of elasticity deduced from a second-gradient of displacement elasticity or strain-gradient elasticity (see Section 9.2). It is a one-dimensional (in space) dispersive but nondissipative nonlinear model deduced from the following Lagrangian and energy densities:

$$L = \frac{1}{2}\, u_t^2 - W(u_x, u_{xx}), \quad W = \frac{1}{2}\left(u_x^2 + \frac{2}{3}\, \varepsilon\, u_x^3 + \varepsilon\, \delta^2\, u_{xx}^2 \right). \quad (10.1)$$

Contrary to the model of nonlinear elasticity selected in Section 8.5 where the elastic energy was given by a quartic, here the elastic potential W presents a cubic contribution with small parameter ε. The weak nonlocality is given by the characteristic length δ (with the *same* small parameter ε in factor). The Lagrangian function possesses all good features to exhibit the typical compensation between dispersion and nonlinearity that is emblematic of solitonic systems.

The resulting Euler–Lagrangian equation of motion is the following nonlinear dispersive wave equation (the celebrated *Boussinesq equation*):

$$u_{tt} - u_{xx}\,(1 + \varepsilon\,u_x) - \varepsilon\,\delta^2\,u_{xxxx} = 0. \tag{10.2}$$

This equation is none other than the field equation (balance of linear physical momentum)

$$\frac{\partial}{\partial t}(u_t) - \frac{\partial}{\partial x}(\sigma^{\text{eff}}) = 0, \tag{10.3}$$

with

$$\sigma^{\text{eff}} = \frac{\delta W}{\delta u_x} = \sigma - m_x, \qquad \sigma = \frac{\partial W}{\partial u_x}, \qquad m = \frac{\partial W}{\partial u_{xx}}. \tag{10.4}$$

The energy equation is easily established, while the balance of field momentum is given by

$$\frac{\partial}{\partial t}(p^w) - \frac{\partial}{\partial x}(b^{\text{eff}}) = 0, \tag{10.5}$$

with

$$p^w = -u_x\,u_t, \qquad b^{\text{eff}} = -(L + u_x\,\sigma + 2\,u_{xx}\,m) + (u_x\,m)_x. \tag{10.6}$$

These are 1D specializations of equations (9.6) and (9.7).

Here, in 1D the volume element representative of a solitary wave motion can only be the whole real line R. By space integration, we obtain the global conservation of wave momentum as

$$\frac{dP(R)}{dt} = \left[b^{\text{eff}}\right]_{-\infty}^{+\infty}, \qquad P(R) := \int_R p^w\,dx. \tag{10.7}$$

This is a Newtonian equation of motion in which the jump between the two limit end values of the effective Eshelby "tensor", here a scalar, plays the role of a *driving force*. A material inhomogeneity or an additional externally prescribed term in the right-hand side of (10.2) will also bring additional driving forces in the right-hand side of $(10.7)_1$. In the absence of such a term, Equation (10.2) possesses strongly localized nonlinear solutions (kinks for u and humps for u_t) with appropriate space and time derivatives vanishing at infinities. Thus equation (10.7) reduces to the equation of an *inertial motion* for these solutions:

$$\frac{dP(R)}{dt} = 0 \text{ and } \frac{dH(R)}{dt} = 0; \qquad H(R) := \int_R H\,dx. \tag{10.8}$$

If an external force perturbing density $\mu\,f(x)$ were acting in the right-hand side of (10.3), μ being a small parameter, then the first of (10.8) would be perturbed in the following way

$$\frac{dP}{dt} = -\varepsilon^2 \int_R u_x\,f(x)\,dx, \tag{10.9}$$

where we took $\mu = O(\varepsilon)$, and one has to find from (10.9) the modulation of the parameters of the localized nonlinear wave due to the perturbation. In order to go further one needs an explicit solution of the unperturbed equation (10.3). Looking for a solution of the plane-wave type $u(x,t) = \tilde{u}(x - ct)$, one is led to possible solutions of the "kink" type:

$$u(x,t) = -\sqrt{1-c^2}\, \tanh\left\{\sqrt{1-c^2}\,(x-ct)\right\}, \qquad (10.10)$$

of which the derivatives vanish at $\pm\infty$, so that the first of (10.8) is satisfied (inertial motion). But the Boussinesq equation has acquired much of its celebrity through the derived equation called the *Korteweg–de Vries (KdV) equation* that we examine now.

10.2.2 *The Korteweg–De Vries equation*

The *Korteweg–de Vries (KdV) equation* is none other than the *one-directional* wave equation — or evolution equation — deduced from (10.2) by means of the so-called *reductive perturbation method* (*cf.* Newell (1985)). The application of this perturbation technique amounts to introducing a right-running characteristic variable $\zeta = x - t$ of the linear part of (10.2) as also a slow time scale $T = \varepsilon T$ and avoiding that the asymptotic expansion

$$u(x,t) = u_1(\zeta,T) + u_2(x,t) + \ldots \qquad (10.11)$$

becomes invalid except if the T-dependence of u_1 is chosen in such a way that the right-hand side of the equation satisfied by u_2 vanishes. This manipulation results in the *secularity condition* written as the equation

$$\frac{\partial q}{\partial \tau} + 6q\,\frac{\partial q}{\partial \tau} + 2\delta^2\,\frac{\partial^3 q}{\partial \zeta^3} = 0, \qquad (10.12)$$

where we have set $q = (1/6)\,\partial f/\partial \zeta$ and $\tau = T/2$. Upon rescaling and changing the notation, this yields the *KdV* equation in the form

$$\frac{\partial v}{\partial t} + v\,\frac{\partial v}{\partial x} + \frac{c_0^2}{2\,k_0^2}\,\frac{\partial^3 v}{\partial x^3} = 0, \qquad (10.13)$$

where c_0 is a typical material velocity and k_0 is a characteristic wave number related to the geometry of the problem (e.g., lattice spacing in elastic crystals). Equation (10.12) admits exact *solitary wave solutions* in the form of a so-call *hump*:

$$v = v_0 \operatorname{sech}^2\left(\frac{x-ct}{\Delta}\right), \qquad (10.14)$$

under the condition that speed c and amplitude v_0 be related by

$$v_0 = 3\,c, \quad \Delta^2 = \frac{2\,c_0}{c\,k_0^2}, \quad \text{hence } v_0\,\Delta^2 = \frac{6\,c_0}{k_0^2} = \text{const.} \tag{10.15}$$

Accordingly, the faster is the wave, the narrower is its profile. We also note that the wavelike solution is *supersonic* as $c > c_0$. Not only does equation (10.13) admit exact localized solutions such as (10.14) but such solutions, although truly nonlinear, in some sense practically superimpose linearly each other since two such solutions traveling in opposite direction interact without further perturbation than a change in phase, recovering their individuality after encounter (collision). This is the property of being **solitonic** *per se* in a strict mathematical sense (in nonexactly solitonic systems, the interactions of "individuals" are usually accompanied by the production of radiation).

As a matter of fact, it was soon realized by the pioneers and creators of soliton theory, e.g., Kruskal and Zabusky (1966); Kruskal (1974) for the *KdV* equation, that systems of equations prone to the pure solitonic type of dynamic behavior admit new conservation laws in addition to the usual ones. It was further shown that for exactly integrable systems (those admitting indeed true soliton solutions), there exists an infinite number of conservation laws and special algorithms were developed to generate these conservations laws (in this regard, see, e.g., Ablowitz and Segur (1981, Section 1.6)). If the field equations used to describe solitons are derived in a field-theoretic context from a Lagrangian (or Hamiltonian) formulation, then these new conservation laws correspond to *symmetry properties* and result from the application of Noether's theorem (Fokas (1979)). However, only a few of these conservation laws can bear an easily grasped physical significance.

The reader may now have realized where we want to lead him since those easily meaningfully interpreted conservation laws are those which pertain to our Eshelbian or canonical framework. These conservation laws are those which are more critically related to the particle-like features of true solitons — which are thus kinds of *quasi-particles* — in the course of so-called elastic collisions or interactions. Canonical (wave) *momentum* is one of these features. Other such features are *mass* and *energy*, these three quantities forming together if possible a true *point mechanics* of which the type will depend on the starting system of partial differential equations. Newtonian and Lorentzian–Einsteinian point mechanics are examples of such mechanics. Others can be created that no direct evidence could bring to the fore.

Remark 10.1. Of course, a duality between soliton-like solutions of some systems issued from quantum physics and elementary particles was rapidly established by nuclear and high-energy physicists (see Rebi and Soliani (1984)). We must also remember the attempts of Louis de Broglie and David Bohm to reconcile quantum physics and a causal interpretation by introducing, in a nonlinear framework, the notion of *pilot wave* guiding the amplitude of the probability ($|\psi|^2$) of presence of a particle as a wave of singularity for which conservation laws and a hydrodynamic analogy play an essential role (see Holland (1993, pp. 113–124), and also Jammer (1974)). It is possible that the present developments bear some relationship to this, but we emphasize that we are mostly interested in macroscopic problems issued from engineering sciences and phenomenological physics (e.g., in the crystalline state).

Remark 10.2. The *KdV* equation (10.13) can also be itself written as a conservation law (here with a different normalization):

$$\frac{\partial v}{\partial t} + \frac{\partial}{\partial x}(3\,v^2 + v_{xx}) = 0. \tag{10.16}$$

A straightforward application of the powerful algorithm proposed by Ablowitz and Segur (1981, p. 56) yields the following next order conservation law

$$\frac{\partial}{\partial t}(v^2 + v_{xx}) + \frac{\partial}{\partial x}(4\,v^3 + 8\,v\,v_x + 5\,v_x^2 + v_{xxxx}) = 0. \tag{10.17}$$

But in early studies of the *KdV* equation (*cf.* Miura (1974)), when such algorithms did not exist, it was proposed to consider the following conservation law:

$$\frac{\partial}{\partial t}\left(\frac{1}{2}\,v^2\right) + \frac{\partial}{\partial x}\left(2\,v^3 + v\,v_{xx} - \frac{1}{2}\,v_x^2\right) = 0. \tag{10.18}$$

It was shown by Maugin and Christov (2002) that the difference between (10.17) and (10.18) is also a conservation law. This shows that there exists an infinity of conservation laws associated with (10.16) and containing the contribution v^2 in the conserved quantity. Here also, a false symmetry between successive conservation laws can be built, being merely an artefact of the one-dimensionality in space of the considered system. For instance one can build a conservation law where the conserved quantity is nothing but the flux present in (10.16) — *cf.* Maugin and Christov (2002). However, the first conserved quantity in (10.16) suggests the introduction of the *potential* $\bar{u}$ of v by $v = \bar{u}_x$, so that we can introduce the conserved mass M_0 by

$$M_0 = \int_R v\,dx = [\bar{u}]_{-\infty}^{+\infty}, \tag{10.19}$$

where $[..]$ denotes the difference (the "jump") between values of the enclosure at the two infinities along the real line (remember that this $\bar{u}$ will be of the *tanh* type). That is, alternatively, M_0 may be called a "difference of potential" or a "voltage" of the solution.

What about the energy? It would seem that v^2 would be a good local measure of the energy. But this is not the case. As a matter of fact, we may consider the following total Hamiltonian:

$$H = -\int_R \left(v^3 - \frac{1}{2}v_x^2\right) dx. \tag{10.20}$$

With $(v, \bar{u})$ playing the role of the Hamiltonian variables usually denoted by the couple (p, q) the first and second Hamilton's equations read

$$q_t = \frac{\delta H}{\delta p} \quad \text{or} \quad \bar{u}_{xt} = v_x, \tag{10.21}$$

and

$$v_t = -\frac{\partial}{\partial x}\left(\frac{\delta H}{\delta v}\right), \tag{10.22}$$

where we introduced a functional derivative. Equation (10.22) is a compatibility condition. Equation (10.22) is none other than the *KdV* equation. On account of the expression (10.20) under the integral sign, we see that the equation of conservation of energy must be the third conservation law as proposed by Bathnagar (1979, p. 126), i.e.,

$$\frac{\partial}{\partial t}\left(v^3 - \frac{1}{2}v_x^2\right) + \frac{\partial}{\partial x}\left(\frac{9}{2}v^4 + 3v^2 v_{xx} + \frac{1}{2}v_x^2 + v_x\,v_t\right) = 0. \tag{10.23}$$

As to the wave momentum, with $v = \bar{u}_x$, the total wave momentum of the v-motion should read

$$P = -\int_R \bar{u}_x\,\bar{u}_t\,dx. \tag{10.24}$$

But we can check that $\bar{u}$ satisfies the following nonlinear evolution equation:

$$\bar{u}_t + 3\,\bar{u}_x^2 + \bar{u}_{xxx} = 0. \tag{10.25}$$

As a consequence, P transforms to

$$P = \int_R (3\,\bar{u}_x^3 + \bar{u}_x\,\bar{u}_{xxx})\,dx = \int_R v\,(3\,v^2 + v_{xx})\,dx, \qquad (10.26)$$

where we recognize the two quantities involved in the conservation law (10.16). All these manipulations suggest that, in so far as the quasi-particle description of the one-directional wave equation is concerned, the basic form is to be found in the original two-directional wave equation, here the Boussinesq equation. This we can show briefly by considering what happens with the so-called "good" or "improved" Boussinesq equation.

Remark 10.3 (The "good" Boussinesq equation). Equation (10.2) has interesting mathematical properties, but it suffers from a serious defect from the point of view of physics: it yields anomalous dispersion for short wavelengths (large wave numbers) in crystals; this corresponds to an instability in Hadamard's sense for partial differential equations. One way to remedy this was to propose a change in sign in factor of the dispersive contribution in (10.2). That is, introducing a new non-dimensionalization, we contemplate the "good" Boussinesq equation in the form

$$u_{tt} - u_{xx} - (u^2 - u_{xx})_{xx} = 0. \qquad (10.27)$$

This can be rewritten as the following Hamiltonian system (see Sanz-Serna and Calvo (1994)):

$$u_t = q_x, \qquad w = u_x, \qquad q_t = w_x^2 + w_{xx} - w, \qquad (10.28)$$

by introducing the auxiliary variables q and w that are in fact defined by the first two of (10.28). The mass M, momentum P, and energy E of soliton solutions of (10.27) or (10.28) are given by

$$M = \int_R u\,dx, \qquad (10.29)$$

$$P(R) = -\int_R u\,q\,dx, \qquad (10.30)$$

and

$$E(R) = \frac{1}{2}\int_R \left(q^2 + w^2 + u^2 + \frac{2}{3}u^3\right)dx. \qquad (10.31)$$

As the system considered is exactly integrable, the quantities just defined are strictly conserved. But their expressions may look somewhat awkward. However, introducing the potential $\bar{u}$ by $u = \bar{u}_x$ with the condition $\bar{u}(x \to -\infty) = 0$, it is verified that

$$M = [\bar{u}]_{-\infty}^{+\infty}, \qquad \bar{u}_t = q, \qquad u\,q = \bar{u}_x\,\bar{u}_t, \qquad \frac{1}{2}q^2 = \frac{1}{2}\bar{u}_t^2, \qquad (10.32)$$

so that M has the same interpretation as in the KdV case, while P and E indeed take their canonical definitions in terms of the potential $\bar{u}$. Simultaneously, in terms of elasticity theory, it is $\bar{u}$ that has the meaning of a displacement while u is a strain *per se*. But accepting the general philosophy of continuum mechanics, we can also consider (10.27) as a field equation issued from second-grade nonlinear elasticity and multiply it by u_x and integrate by parts to arrive at the equation of wave momentum (*cf.* Eq. (10.5))

$$\frac{\partial}{\partial t} p^w - \frac{\partial}{\partial x} b^{\text{eff}} = 0, \qquad p^w := -u_x u_t. \tag{10.33}$$

10.2.3   *The sine-Gordon equation*

This is the one-dimensional (in space) partial differential equation

$$u_{tt} - u_{xx} + \sin u = 0, \tag{10.34}$$

where both nonlinearity and dispersion are contained in the *sin* function. By introducing right- and left-running characteristic coordinates ξ, $\zeta = x \pm t$, this ubiquitous equation can also be written as *Enneper's equation* of surface geometry as

$$u_{\xi\zeta} - \sin u = 0. \tag{10.35}$$

Equation (10.34) or (10.35) appears not only in the geometry of surfaces with constant negative curvature but also in many fields of physics, especially while studying the structure of magnetic domain walls and Josephson junctions. From the point of view expanded in this volume, such an equation can be formulated as an elementary model of dislocation motion in the so-called Frenkel–Kontorova (1938) model: a linear atomic chain placed in a sinusoidal potential landscape. The name is granted by analogy with the Klein–Gordon equation of atomic physics (where $\sin u$ is simply replaced by u).

This remarkable equation is *exactly integrable* (i.e., admits true soliton solutions) and is *Lorentz invariant*. It admits single subsonic solitary-wave solutions of the following *kink* form

$$u(x,t) = \bar{u}(\xi) = 4 \tan^{-1} \left[\exp \left(\pm \gamma \left(\xi - \xi_0 \right) \right) \right], \tag{10.36}$$

where we have set

$$\xi = x - ct, \qquad \gamma = (1 - c^2)^{-1/2}, \qquad |c| < 1. \tag{10.37}$$

Considered as an *elastic* system, Equation (10.34) is derivable from the following Lagrangian–Hamiltonian framework where the sinusoidal term should be interpreted as the action of an external source (e.g., the already mentioned periodic substrate of the Frenkel–Kontorova modelling) since the classical elastic energy cannot depend explicitly on u:

$$L = \frac{1}{2}\,u_t^2 - \frac{1}{2}\,u_x^2 - (1 - \cos u), \qquad p = \frac{\partial L}{\partial u_t}, \tag{10.38}$$

$$H = p\,u_t - L = \frac{1}{2}\,(p^2 + u_x^2) + 2\,\sin^2(u/2), \tag{10.39}$$

with Hamiltonian equations

$$u_t = \frac{\partial H}{\partial p} = p, \qquad p_t = -\frac{\delta H}{\delta u} = u_{xx} - \sin u. \tag{10.40}$$

A kink ($2\,\pi$ solution in u) or an anti-kink ($-2\,\pi$ solution) may be considered as a quasi-particle with rest mass M_0 — here prescribed as a pure number —, momentum P, and energy E given by

$$M_0 = 8 = E(0), \tag{10.41}$$

$$P = P(R) = \int_R (-u_x\,u_t)\,dx = 8\,\gamma\,c = M\,c = \frac{M_0\,c}{\sqrt{1 - c^2}}, \tag{10.42}$$

$$E = E(R) = \int_R H\,dx = 8\,\gamma = E(c), \tag{10.43}$$

with the classical relationship between the triplet (M_0, P, E) typical of *Lorentzian* point-mechanics (the characteristic speed of relativity, the light velocity in vacuum, here is one):

$$E^2(c) = M_0^2 + P^2(c), \tag{10.44}$$

while for Newtonian point mechanics we would classically have

$$E = P^2/2\,M_0, \qquad P = M_0\,c. \tag{10.45}$$

Expressions (10.42) and (10.44) agree with the canonical definitions of Chapter 3. The given estimates are based on the form (10.36) of the kink solutions. Note that here the amplitude of the solution is independent of the speed: we have a so-called *topological* soliton. Because of this the solutions (10.36) exist also in *statics* and can, therefore, represent the structure of a magnetic domain wall at rest as originally found by Landau and Lifshitz. Of course, for a fixed nonvanishing c, the three quantities M, P, and E are strictly conserved, i.e., for kink and antikink solutions; in particular, we have a standard *inertial* equation of motion for a relativistic point quasi-particle:

$$\frac{dP}{dt} = \frac{d}{dt}\left(\frac{M_0\,c}{\sqrt{1 - c^2}}\right) = 0. \tag{10.46}$$

10.2.4   *The nonlinear Schrödinger model*

Another well known exactly integrable equation (*cf.* Calogero and De-gasperis (1982)) is the cubic Schrödinger equation (sometimes called *NLS* without specializing to the cubic case). In a one-dimensional spatial setting, this equation reads:

$$i\,a_t + a_{xx} + 2\,\lambda\,|a|^2\,a = 0, \tag{10.47}$$

where $a(x,t)$ is a complex-valued amplitude and λ is a real scalar parameter. This is not directly related to a problem from the elasticity of crystals. But it is of somewhat universal application as it governs the slowly varying complex amplitude of the envelope of a carrier wave (fast oscillations) in a *dispersive, weakly nonlinear* medium, of which a special case may indeed be an elastic crystal (See Section below). This equation admits so-called bright and dark solitons as solutions (of which the exact form here is quite irrelevant; see Maugin *et al.* (1992a), for these) for which, to start with, the mass M, canonical momentum $P(R)$, and energy $E(R)$ are given by (*cf.* Drazin and Johnson (1989))

$$M = \int_R |a|^2\,dx, \tag{10.48}$$

$$P(R) = \int_R i\,(a\,a_t^* - a^*\,a_t)\,dx, \tag{10.49}$$

$$E(R) = \int_R \frac{1}{2}\,(|a_x|^* - \lambda\,|a|^4)\,dx, \tag{10.50}$$

where the asterisk indicates the complex conjugate. The quantum physicist will recognize in (10.48) the total probability of presence of a particle of wave function a according to Max Born's interpretation (this should be normalized to one). In our mechanical frame of mind, this, without normalization, may be called the *number of surface phonons* or, also the *wave action*. The canonical momentum P was introduced in the causal re-interpretation of quantum mechanics (see Holland (1993, p. 113)) by treating a as a classical but complex-valued field. Remarkably, the point mechanics associated with (10.48) to (10.50) is Newtonian; that is, if c is the propagation velocity of bright solitons, we have the following Newtonian relations among the triplet (M, P, E) [Compare to (10.45)]

$$P = M\,c, \qquad E = P^2/2\,M. \tag{10.51}$$

This is shown by substituting the bright-soliton solution obtained by Za-kharov and Shabat (1972) in the set (10.48)–(10.50).

10.2.5 Comments

Here above we have briefly evoked four cases of often met model equations. Since these are more models than true explicit field equations of specific problems of acoustic physics, they must be considered as ideals that will rarely be met in their precise form and simplicity when attacking true practical problems. In many real applications what is truly found are generalizations of these model equations with perturbing terms or couplings with other equations that hinder the exact integrability. For instance, a further extension of the gradient model of elasticity will generate field equations that are more complex than the traditional Boussinesq equation. Such cases yield strange behaviours for the associated quasi-particles as illustrated in Section 10.3. The usual Korteweg–de Vries equation may be generalized to equations having higher power terms in both the nonlinear and dispersive contents. This may give rise to systems admitting soliton solutions having a compact support, so-called compactons (*cf.* Rosenau and Hyman (1993)).

In some other physical problems, the sine-Gordon equation is coupled to another wave equation. This occurs in the magneto-elasticity of ferromagnets, where the coupling is due to magnetostriction. Similar effects can be expected in elastic ferroelectric crystals where the dynamics of electric dipoles is coupled to the deformation *via* electrostriction. These two cases have been studied in detail by one of the authors and some co-workers (regarding this point, see Maugin (1988, Chap. 6); Maugin *et al.* (1992a); Maugin (1999b)). This will not be documented here although the resulting systems have been nicknamed *sine-Gordon–d'Alembert systems* by Kivshar and Malomed (1989). As to the nonlinear Schrödinger equation we shall show in Section 10.4 how such an equation can result from an analysis of surface nonlinear elastic waves that improves on the model introduced by Murdoch and recalled in Section 7.3. To deal with such a case requires considering some of the ideas and techniques introduced in Chapter 4.

10.3 The generalized Boussinesq model (gradient elasticity)

There are several ways to generalize the Boussinesq equation. These generalizations will obviously lead to nonexactly integrable systems. One such system is obtained while studying the ferro-elastic phase transition as a dynamical process in elastic crystals. Here we illustrate our purpose with a modelling proposed by Christov and Maugin (1993) when approaching the difference scheme of lattice dynamics in a more accurate way than usually done. With $s = v_x$ a *shear strain*, from a lattice-dynamics approach and a long-wavelength limit while neglecting coupling with other strain components, one obtains the following type of equation:

$$s_{tt} - c_T^2 \, s_{xx} - \left(F(s) - \beta \, s_{xx} + s_{xxxx} \right)_{xx} = 0. \qquad (10.52)$$

Here F is a polynomial in s starting with second degree (it could even be a nonconvex function admitting three minima), c_T is a characteristic speed and β is a positive scalar. It can be said that both the nonlinearity and dispersion have been increased compared to the classical Boussinesq equation. Equation (10.52) is *stiff* in the sense that it involves a sixth-order space derivative, a situation that obviously imposes rather strong limit conditions at infinity or at the ends of a finite interval in numerical simulations. In spite of its apparent complexity (10.52) admits solitary-wave solutions which involve the ubiquitous *sech* function (but at the fourth power) for a single value of the phase speed — the existence of different solitary-wave solutions with a continuous spectrum for c was shown numerically (Christov and Maugin (1993)). But it is true that for a velocity too close to c_T these solutions are not able to preserve their shape and eventually they transform into pulses. The latter, in turn, exhibit a self-similar behaviour as long as the amplitude of the pulses decreases while their support increases (a phenomenon analogous to a *red shift*). These pulses practically pass through each other without changing qualitatively their shapes — save the red-shifting — with perfect conservation of "mass" and "energy", so that theses pulses may qualitatively be claimed to be "solitons".

The search for the properties of associated quasi-particle motions leads to original results. The "mass", "momentum" and "energy" of the system (10.52) can be defined thus. Let $F(s) = -dU(s)/ds$. First, following along the same line as what done for (10.27), we rewrite (10.52) as a *Hamiltonian system* by introducing the triplet (s, q, w) such that (10.52) is equivalent to the following system:

$$s_t = q_{xx}, \qquad w = s_{xx}, \qquad q_t = c_T^2 \, s + F(s) - \beta \, w + w_{xx}. \qquad (10.53)$$

Then we have

$$M = \int_R s \, dx = \left[v \right]_{-\infty}^{+\infty}, \qquad (10.54)$$

$$P(R) = -\int_R s \, q_x \, dx = -\int_R v_x \, v_t \, dx, \qquad (10.55)$$

$$E(R) = \frac{1}{2} \int_R \left(c_T^2 \, s^2 + q_x^2 - 2 \, U(s) + \beta \, s_x^2 + w^2 \right) dx, \qquad (10.56)$$

i.e.

$$E(R) = \frac{1}{2} \int_R \left(v_t^2 + c_T^2\, s^2 - 2\, U(s) + \beta\, s_x^2 + s_{xx}^2 \right) dx. \tag{10.57}$$

Here we have assumed that $v_t(-\infty) = 0$, so that the transformations indicated in the second of (10.55) and in (10.57) — canonical definitions for wave momentum and (kinetic plus potential) energy — are valid. Then there hold the following global balance laws:

$$\frac{dM}{dt} = 0, \qquad \frac{dE}{dt} = 0, \tag{10.58}$$

and

$$\frac{d}{dt} P(R) = \overline{F} := \left[s_{xx}^2 \right]_{-\infty}^{+\infty}. \tag{10.59}$$

For solitary-wave solutions, the driving force $\overline{F}$ is in fact equal to zero by virtue of the asymptotic conditions. That force $\overline{F}$, for a long but finite interval of simulation on R, is felt only when the solitons "hit" the boundaries and rebound from them. Yet the energy remains unchanged. A strongly implicit conservative finite-difference scheme is used in numerical simulations in order to always preserve both M and E. More on the quasi-particle mechanics corresponding to (10.54) through (10.56) is to be found in Christov *et al.* (1996). The exhibited particle-like behaviour is dominated by an *anti-Lorentzian character* (i.e., while mass varies with the speed, all kinetic quantities go to zero — and not infinity — at a critical speed). A numerical fit has allowed these authors to uncover this new "point-mechanics" which reduces to a Newtonian one for small speeds, i.e., then $M \approx M_0$ and $P \approx M_0\, c$ for a definite M_0. This is true for solutions of the monotonous *sech*-like shapes (solutions) and also so-called *Kawahara* solitons (localized soliton hump-like solutions which acquire oscillatory tails on both sides — so-called "*nanopterons*").

Indeed, since there are no known analytical expressions for mass, wave-momentum and energy in our specific problem, it is found by best numerical fit that for sech-like solutions, one obtains (only powers of γ that are integer multiples of $1/2$, $1/3$, or $1/4$ are hypothesized)

$$M = M_0\, \gamma^{5/4}, \quad P = M\, c\, \gamma^2 = M_0\, c\, \gamma^{13/4} \quad \text{for } M_0 = 7.4. \tag{10.60}$$

It was not possible to find a reasonable expression for the energy for this case where the celerity of the localized waves ranges from zero to the characteristic velocity c_0 (here equal to unity). Here,

$$\gamma = \{ 1 - (c^2/c_0^2) \}^{1/2} = \sqrt{1 - c^2}.$$

For the Kawahara type of solitons, in parallel with (10.60) one obtains

$$M = M_0 \gamma^{7/4}, \quad P = M c \gamma^{5/4} = M_0 c \gamma^{12/4} \quad \text{for } M_0 = 2.986. \quad (10.61)$$

In both cases (10.60) and (10.61), we see that the "relativistic Lorentz" parameter γ intervenes at a positive power, so that this justifies the "anti-Lorentzian" denomination for the corresponding behaviour of the associated quasi-particles. But both admits the Newtonian limit $M \approx M_0$ and $P \approx M_0 c$ for $c \ll c_0$. We refer the reader to Maugin and Christov (2002) for more detail.

As a final remark, we note that it is difficult to imagine before hand what will be the "point mechanics" of quasi-particles associated with unusual systems of dynamical field equations. The case of the sine-Gordon equation examined in Paragraph 10.2.3 was exceptional in that the starting field equation was already Lorentz-invariant. But the case examined in Paragraph 10.2.4 was also surprising by the very simplicity of the Newtonian behaviour encapsulated in Equations (10.51) without any approximation. We move next to a generalization of this situation by examining a surface wave acoustic problem.

10.4 Surface elastic solitons

10.4.1 *The basic equations*

In this section we combine ideas from Section 7.3 and Section 8.5 by examining the possibility of obtaining surface solitons guided by an elastic structure. Indeed, we assume like in Section 7.3 that a *linear elastic film* is perfectly glued on top of the surface (this will introduce the necessary dispersion) and, in agreement with Section 8.5, that the substrate is a *nonlinear elastic body* (this involves the requested nonlinearity). Furthermore, like for the Murdoch surface wave, there is no need of electric properties here, and both the film and the substrate are assumed to be isotropic and homogeneous for the sake of simplicity. The main equations are as follows.

In an obvious index Cartesian tensor notation, we consider the following surface-wave problem within the Murdoch model of elastic bodies (to which we will add both physical and geometric nonlinearities in the substrate) with a material boundary.

First we recall Equations (7.97) and (7.98):

- For $x_2 > 0$ (semi-infinite elastic space)

$$\rho_0 \frac{\partial^2 u_i}{\partial t^2} = \frac{\partial}{\partial x_j} \sigma_{ji}; \tag{10.62}$$

- For $x_2 < 0$ we have a vacuum.
- At $x_2 = 0$ (material surface with mass, inertia, and elasticity)

$$\hat{\rho}_0 \frac{\partial^2 \hat{u}_i}{\partial t^2} = \frac{\partial}{\partial \hat{x}_j} \hat{\sigma}_{ji} - n_j \, \sigma_{ji}^+. \tag{10.63}$$

Here superimposed carets refer to quantities related to the surface of unit outward oriented normal n_j (directed towards negative x_2's). Thus $\hat{\sigma}_{ji}$ is a surface stress while $\sigma_{ji}^+ = \lim \sigma_{ji}$, $x_2 \to 0^+$, is the three-dimensional stress from the body. Mass density $\hat{\rho}_0$ is per unit surface.

With a wave problem depending only on the coordinates (x_1, x_2) in the sagittal plane Π_S and a purely SH displacement u_3 in the x_3 direction orthogonal to Π_S, the relevant components of equations (10.62) and (10.63) are formally given by Equations (7.99) and (7.100). That is,

$$\rho_0 \frac{\partial^2 u_3}{\partial t^2} = \frac{\partial}{\partial x_1} \sigma_{13} + \frac{\partial}{\partial x_2} \sigma_{23}, \qquad x_2 > 0, \tag{10.64}$$

and

$$\hat{\rho}_0 \frac{\partial^2 \hat{u}_3}{\partial t^2} = \frac{\partial}{\partial \hat{x}_1} \hat{\sigma}_{13} + \sigma_{23}, \qquad x_2 = 0. \tag{10.65}$$

The stress components are given by

$$\sigma_{ji} = \frac{\partial W}{\partial e_{ji}}, \qquad \hat{\sigma}_{ji} = \frac{\partial \widehat{W}}{\partial \hat{e}_{ji}}, \tag{10.66}$$

where, in the nonlinear case, the bulk elastic energy is given for an isotropic body by Murnaghan's expression (*cf.* Bland (1969); Murnaghan (1951))

$$W = \frac{\lambda}{2} I_1^2 + \mu I_2 + \alpha I_1^3 + \bar{\beta} I_1 I_2 + \gamma I_3 + \xi I_1^4 + \nu I_1 I_3 + \delta I_2^2, \tag{10.67}$$

$$I_\alpha = \text{trace } \mathbf{e}^\alpha, \tag{10.68}$$

while the surface energy considered only as quadratic for an "isotropic" surface is given by

$$\widehat{W} = \frac{\widehat{\lambda}}{2} \hat{I}_1^2 + \hat{\mu} \, \hat{I}_2, \qquad \hat{I}_\alpha = \text{trace } \hat{\mathbf{e}}^\alpha. \tag{10.69}$$

All computations done, we obtain thus the set of two equations:

$$\frac{\partial^2 u_3}{\partial t^2} = c_T^2 \frac{\partial}{\partial x_1} \big[u_{3,1} \left(1 + \bar{\delta}_{\text{eff}} \, \Phi \right) \big]$$

$$+ c_T^2 \frac{\partial}{\partial x_2} \big[u_{3,2} \left(1 + \bar{\delta}_{\text{eff}} \, \Phi \right) \big], \qquad x_2 > 0, \tag{10.70}$$

and

$$\frac{\partial^2 \hat{u}_3}{\partial t^2} = \hat{c}_T^2 \frac{\partial^2 \hat{u}_3}{\partial x_1^2} + \hat{c}_T^2 \, k_\alpha \, u_{3,2} \left(1 + \bar{\delta}_{\text{eff}} \, \Phi \right), \quad x_2 = 0, \tag{10.71}$$

where we have set

$$c_T^2 = \frac{\mu}{\rho_0}, \qquad \hat{c}_T^2 = \frac{\hat{\mu}}{\hat{\rho}_0}, \qquad k_\alpha = \frac{\mu}{\hat{\mu}}, \tag{10.72}$$

$$\bar{\delta}_{\text{eff}} = \frac{\delta_{\text{eff}}}{\mu}, \qquad \delta_{\text{eff}} = \delta + \left(\bar{\beta} + \frac{3}{2} \gamma \right) + \left(\frac{\lambda}{2} + \mu \right), \tag{10.73}$$

and the typical strain invariant

$$\Phi = u_{3,1}^2 + u_{3,2}^2. \tag{10.74}$$

Here k_α is a characteristic wave number (inverse of a characteristic length), and δ_{eff} is an effective nonlinear elasticity coefficient that accounts for elasticities of the first, second, third and fourth order.

Remark 10.4. The above condensed presentation may be misleading because, in truth, equations (10.70) through (10.73) are obtained from the initial consideration of a Lagrangian formulation where σ would be the second (symmetric) Piola–Kirchhoff stress and e a Lagrangian strain (See Chapter 2). This explains the definition of the effective coefficient in the second of (10.73) when we pass to the displacement gradient, a result typical of the combination of both physical *and* geometrical nonlinearities in the present context (proof of this in Maugin and Hadouaj (1991); all details in the Ph. D Thesis of H. Hadouaj, University of Paris-6, Paris, 1991).

Furthermore, we must check the condition of perfect gluing of the material interface on the body

$$u_3(x_1, x_2 = 0; t) = \hat{u}(x_1; t), \tag{10.75}$$

and the radiation condition far from the surface:

$$u_3(x_1, x_2 \to \infty; t) = 0. \tag{10.76}$$

To revert to the original work of Maugin and Hadouaj (1991), we introduce their notation:

$$\Delta = c^2 \, \Delta_{\text{eff}}, \qquad \Delta_{\text{eff}} = \frac{\delta_{\text{eff}}}{\mu}, \qquad \beta = \frac{c_S}{c_T}, \tag{10.77}$$

$$c_S^2 \equiv \hat{c}_T^2 = \frac{\hat{\mu}}{\hat{\rho}_0}, \quad c_T^2 = \frac{\mu}{\rho_0}, \quad X = k_\alpha \, x_1, \; Y = k_\alpha \, x_2, \; U = k_\alpha \, u_3.$$

Then it is shown that Equations (10.70) and (10.71) can be rewritten in the following form:

$$\text{Dal}_B U \equiv \beta^2 U_{TT} - (U_{XX} + U_{YY}) = \beta^2 \, \Delta \, T_B^{NL}(U) \quad \text{for } Y > 0, \quad (10.78)$$

$$\text{Dal}_S U \equiv \widehat{U}_{TT} - \widehat{U}_{YY} = U_Y + \beta^2 \, \Delta \, T_S^{NL}(U) \quad \text{for } Y = 0, \quad (10.79)$$

with

$$U = \widehat{U} \text{ at } Y = 0, \qquad U(X, Y \to \infty, T) = 0, \quad (10.80)$$

and the following definitions for the source terms:

$$T_B^{NL} = (U_X \, \Phi)_X + (U_Y \, \Phi)_Y, \quad T_S^{NL} = U_Y \, \Phi, \quad \Phi := U_X^2 + U_Y^2. \quad (10.81)$$

The abbreviation *Dal* means a Dalembertian operator. Times are normalized with respect to a characteristic frequency ω_a so that $c_S^2 = \omega_a^2/k_a^2$. The scalar Δ now stands for the effective nonlinearity parameter. It may be of any sign depending on the material of the substrate; $\beta < 1$ is the dispersion parameter.

Remark 10.5. The system (10.78) through (10.80) admits the following Lagrangian per unit length in the X direction [*cf.* equation (7.109)]:

$$L = \int_0^\infty \left(\frac{\beta^2}{2} U_T^2 - \frac{1}{2} \Phi - \frac{\Delta \beta^2}{4} \Phi^2 \right) dY + \frac{1}{2} \left(\widehat{U}_T^2 - \widehat{U}_X^2 \right). \quad (10.82)$$

This shows that the effective physical nonlinearity is indeed in a fourth-order elasticity.

10.4.2 *Reminder: Linear harmonic approximation*

This is a solution of the Murdoch type (7.54):

$$U(X, Y, T) = A_0 \exp(-\chi Y) \exp[i (k X - \omega T)]. \quad (10.83)$$

The resulting linear "bulk" and "surface" conditions yield

$$D_B(k, \omega; \chi) \equiv \beta^2 \omega^2 - k^2 + \chi^2 = 0, \quad (10.84)$$

$$D_S(k, \omega; \chi) \equiv \omega^2 - k^2 - \chi = 0. \quad (10.85)$$

Eliminating χ between these two conditions leads to Murdoch's dispersion relation (7.114) or (7.116) in the following disguise:

$$D_L(\omega, k) = D_M(\omega, k) \equiv \beta^2 \omega^2 - k^2 + (\omega^2 - k^2)^2 = 0. \quad (10.86)$$

The existence of such a surface dispersive monomode requires that $c_S < c_T$ or $\beta < 1$ (*cf.* Section 7.4.2).

10.4.3 *Solitary-wave solutions for envelope signals*

A. Kinematic-wave approach

In the nonlinear regime and small amplitudes we can consider solutions of the type

$$U = A \exp(-\chi Y) \cos\theta + \dots, \qquad \theta = k X - \omega T. \qquad (10.87)$$

This implies that the source terms (10.81) — that result from effective fourth-order elasticity — act as *third-harmonic generators*:

$$T_B^{NL} \cong (9\chi^4 - 3k^4 + 2k^2\chi^2) \frac{A^3}{4} \exp(-3\chi Y) \cos\theta$$
$$+ \text{terms in } \cos(3\theta), \qquad (10.88)$$

$$T_S^{NL} \cong -(\chi k^2 + 3\chi^3) \frac{A^3}{4} \cos\theta + \text{terms in } \cos(3\theta). \qquad (10.89)$$

[Note the first factor within parentheses in (10.88) and the same factor in (8.89)].

Now for the amplitude A depending on X, Y and T, this allows one to look for the "nonlinear bulk and surface dispersion relations" in the form of perturbations such as

$$D_B^{NL}(k, \omega, \chi, A)$$
$$\equiv D_B(k, \omega; \chi) + \frac{\Delta\beta^2}{4} (9\chi^4 - 3k^4 + 2k^2\chi^2) A^2 \exp(-2\chi Y) \qquad (10.90)$$
$$= \varepsilon\, l^{(1)} + \varepsilon^2\, l^{(2)} + \dots,$$

$$D_S^{NL}(k, \omega, \chi, A) \equiv D_S(k, \omega; \chi) - \frac{\Delta\beta^2}{4} (\chi k^2 + 3\chi^3) A^2 \qquad (10.91)$$
$$= \varepsilon\, m^{(1)} + \varepsilon^2\, m^{(2)} + \dots,$$

in an asymptotic expansion effected by writing

$$x = \varepsilon X, \quad y = \varepsilon Y, \quad t = \varepsilon T; \qquad (10.92)$$
$$U(X, Y, T) = f(\theta, A) + \varepsilon\, U_1 + \varepsilon^2\, U_2 + \dots,$$

where θ is a general phase such that [*cf.* Equations (4.29)]

$$k = \theta_X, \quad \omega = -\theta_T, \quad i\chi = \theta_Y. \qquad (10.93)$$

On evaluating the partial derivatives $\partial/\partial T$, $\partial/\partial X$, *etc.*, it is found that

$$U_0 = A \exp\left[i\,(k X - \omega T + i\chi Y)\right], \qquad (10.94)$$

which is none other than the linear SAW solution, while at order two one obtains (Maugin and Hadouaj (1991))

$$U_1 = 0, \qquad l^{(1)} = 0, \qquad m^{(1)} = 0, \tag{10.95}$$

together with the following equations of *conservation of wave action* (*cf.* Equations (10.96) and (10.97); also Benney and Newell (1967)):

$$\frac{\partial}{\partial t}(\beta^2 \omega A^2) + \frac{\partial}{\partial x}(k A^2) + \frac{\partial}{\partial y}(i \chi A^2) = 0, \qquad y > 0, \tag{10.96}$$

and

$$\frac{\partial}{\partial t}(\omega A^2) + \frac{\partial}{\partial x}(k A^2) - \frac{\partial}{\partial y}\left(\frac{i A^2}{2}\right) = 0 \quad \text{at } y = 0. \tag{10.97}$$

At the next order — applying the secularity conditions — one obtains (10.90) and (10.91) in the definite form

$$D_B^{NL}(\omega, k, \chi, A) - \frac{\varepsilon^2}{A} \mathrm{Dal}_B A = 0, \qquad y > 0; \tag{10.98}$$

and

$$D_S^{NL}(\omega, k, \chi, A) - \frac{\varepsilon^2}{A} \mathrm{Dal}_S A = 0 \quad \text{at } y = 0. \tag{10.99}$$

If we dare say, these are "nonlinear dispersive" dispersion relations that result from a double expansion in which both ε and A are of the same order. Equations (10.98) and (10.99) themselves have become wave equations! To these we must add the conditions (10.92) of kinematic-wave theory in the following form:

$$\frac{\partial k}{\partial t} + \frac{\partial \omega}{\partial x} = 0, \qquad \frac{\partial \chi}{\partial t} - i\frac{\partial \omega}{\partial y} = 0. \tag{10.100}$$

Once χ is eliminated between (10.96), (10.97), (10.98) and (10.99), then the resulting equations and (10.100) are the equations that govern k, ω and A. The solution of this system is in general not manageable except with an additional assumption (see next paragraph). What we just presented sketchily is the essence of the Whitham (1974)–Newell (1985) method. Here it is exceptionally applied to the case of a two- dimensional space (with a 1D true one propagation space and an orthogonal space — in depth in the substrate — that is not propagative).

B. Almost monochromatic limit

Let (ω_0, k_0, χ_0) the wave characteristics of an admissible harmonic regime. Assume that we are close to this by considering infinitesimally small deviations such that

$$k = k_0 + \varepsilon\,\phi_x, \qquad \omega = \omega_0 - \varepsilon\,\phi_t, \qquad \chi = \chi_0 - i\,\varepsilon\,\phi_y, \qquad (10.101)$$

where ϕ is a perturbation phase. Both ϕ and the amplitude A are assumed to vary slowly. We can thus introduce a new phase, time scale and a rescaling of the amplitude by

$$\xi = x - \omega_0'\,t + i\,\chi_0'\,y, \qquad \tau = \varepsilon\,t, \qquad A \to \varepsilon\,A, \qquad (10.102)$$

where primes denote differentiation with respect to k at k_0. Then by a lengthy manipulation that we do not reproduce (see Maugin and Hadouaj (1991)), which amounts to using Equations (10.96) through (10.99) and the first two differentials of Equations (10.84)–(10.85) evaluated at (ω_0, k_0), and finally introducing the complex amplitude $a = A\exp(i\,\phi)$, one is led to a system of two nonlinear Schrödinger (NLS) equations. By eliminating χ_0'' between them at $y = 0$, one shows that the whole problem reduces to the following NLS equation:

$$i\,a_\tau + p\,a_{\xi\xi} + q\,|a|^2\,a = 0, \qquad (10.103)$$

where

$$\tau = \varepsilon\,t, \qquad \xi = x - \omega_0' \qquad y = 0,$$

and

$$p = \frac{1}{2}\,\omega_0'', \qquad q = \frac{3}{8}\,\Delta\,\beta^2\,\frac{\beta^2\,\omega_0^2\,(\beta^2\,\omega_0^2 - 2\,k_0^2)}{\omega_0\,\left[\beta^2 + 2\,(\omega_0^2 - k_0^2)\right]}, \qquad (10.104)$$

where the point (ω_0, k_0) belongs to the linear dispersion relation (10.86).

The parameter p account for the group-velocity dispersion whereas q clearly accounts for nonlinearity.

- For $\Delta > 0$, the positiveness of the product $p\,q$ corresponds to $1 > \beta^2 > 1/2$, and according to Zakharov and Shabat (1972), the NLS equation (10.103) admits so-called stable *bright-envelope solitons* with expression

$$\widehat{U}(X, T) = \varepsilon\,\eta\,\left|\frac{\omega_0''}{q}\right|^{1/2} \exp\left[i\,(k_0\,X - \omega_0\,T + 2\,\varepsilon^2\,\omega_0''\,T\,\eta^2 + \phi_0)\right]$$

$$\operatorname{sech}\left[2\,\varepsilon\,\eta\,(X - \omega_0'\,T - X_0)\right].$$

$$(10.105)$$

This dynamic solution contains three parameters η, X_0 and ϕ_0.

- For $p\,q < 0$, which corresponds to $0 < \beta^2 < 1/2$, instead of (10.105) one obtains so-called *dark-envelope solitons* with expression

$$\widehat{U}(X,T) = \varepsilon \left| \frac{\omega_0''}{q} \right|^{1/2} \tanh\left[\varepsilon\left(X - \omega_0' T - X_0\right)\right] \exp\left[i\left(k_0\, X - \omega_0\, T\right)\right],$$
$$(10.106)$$

which contains only two parameters ε and X_0.

For both solutions (10.104) and (10.105), the dependence on depth is simply obtained by multiplying by $\exp(-\chi_0\, Y)$ with $\chi_0 \cong \omega_0^2 - k_0^2$ just like in the linear solution (*cf.* (10.85)).

Maugin and Hadouaj (1991) have shown that solution (10.105) prevails for substrate made of lithium niobate $LiNbO_3$ and a film of aluminium. Solution (10.106) prevails for a gold or silver thin film glued on the $LiNbO_3$ substrate.

C.   *Quasi-particle behaviour*

Equation (10.103) is exactly integrable by the inverse-scattering method. As shown in Paragraph 10.2.4 it admits a Newtonian point mechanics for its associated quasi-particles. The most characteristic quantity in this mechanics is the mass M given by Equation (10.48), i.e.,

$$M = \int_R |a|^2\, dx. \qquad (10.107)$$

Here this may be referred to as the total *wave action* or the number of surface phonons. But, after some transformations and asymptotics, one may ask the legitimate question whether the original physical system of partial differential equations — system (10.70)–(10.71) with the conditions (10.75)–(10.76) — itself presents the same solitonic behaviour even approximately. The answer is yes as proved by numerical experiments of Hadouaj and Maugin (1992) performed directly on this system with possible injected solutions of the type (10.105) or (10.106). The stable propagation of a single solution (10.105) — with accompanying decrease of amplitude with depth in the substrate — was checked for appropriate values of the parameters while unstable propagation of solution (10.106) was exhibited for the same values of these parameters. The practically solitonic (head-on) collision of two unequal counter running solutions was remarkably shown for sufficiently small amplitudes (see Figure 9.11 in Maugin (1999b)). All this to say that the physical system, in spite of its apparent complexity, is practically solitonic exhibiting only small spurious radiations.

More complex problems involving the present surface solitons have been dealt with in which the SH component of the surface waves is coupled with the remaining Rayleigh component polarized in the sagittal plane. This gives rise to a nonlinear dispersive wave system called the *generalized Zakharov system* (the original Zakharov system appeared in problems dealing with Langmuir ion-acoustic waves in plasmas; Zakharov and Shabat (1972)). A thorough study of this generalized Zakharov system was offered by Hadouaj, Malomed and Maugin in a series of papers that dutifully exploit the perturbed point-mechanics of associated quasi-particles with interesting and sometimes surprising behaviours (see Maugin *et al.* (1992a); Hadouaj *et al.* (1991a,b); also Maugin (1999b, Section 9.5)). Surface solitary longitudinal waves in an elastic crystal with third-order elasticity are examined by Porubov and Maugin (2009).

Appendix A

Appendix to Chapter 3: Reminder on Noether's theorem

Note. This appendix is essentially extracted from Maugin (2011b, Chapter 4).

Here we are concerned with simple general features of field theories in a continuum with space-time parametrization $\{\mathbf{X}, t\}$. We consider *Hamiltonian actions* of the type

$$A(\phi; V) = \int_{V \times T} L\left(\phi^\alpha, \partial_\mu \phi^\alpha, \partial_\mu \partial_\nu \phi^\alpha, \dots; X^\mu\right) d^4 X, \qquad (A.1)$$

where ϕ^α, $\alpha = 1, 2, \dots, N$, denotes the ordered array of fields, say the independent components of a certain geometric object, and $d^4 X = dV\, dt$; V is a material volume and T a time interval. This is a Cartesian–Newtonian notation, with

$$\begin{aligned} &\{\partial_\mu = \partial/\partial X^\mu; \ \mu = 1, 2, 3, 4\} \\ &= \{\partial/\partial X^K; \ K = 1, 2, 3; \ \partial/\partial X^4 = \partial/\partial t\}. \end{aligned} \qquad (A.2)$$

The summation over dummy indices (Einstein convention) is enforced.

We limit the considerations to a so-called first-order gradient theory where the dependency of the Lagrangian density L is limited to the first space-time gradient $\partial_\mu \phi^\alpha$.

From expression (A.1) we can derive two types of equations: those relating to *each one* of the fields ϕ^α, and those which express a general conservation law of the system governing *all* fields simultaneously. The first group is obtained by imposing the requirement that the variation of the action A be zero when we perform a *small* variation $\delta\phi^\alpha$ of the corresponding field under well specified conditions at the boundary ∂V, of V (if V is not the whole of space), and at the end points of the time interval $T = [t_0, t_1]$ if such limitations are considered. However, most field theories are developed for an infinite domain. The second group of equations is the result of the variation of the parametrization, and these results, on account of the former group, express the *invariance* or lack of invariance of the whole system under changes of this parametrization. To simplify the presentation we assume an infinite domain V with vanishing fields at infinity, and an infinite time interval since our concern here is neither boundary conditions, nor initial conditions.

In what follows it is understood that "variation" means "infinitesimal variation", and the symmetries involved are generated by infinitesimally small variations and parameters. In order to perform these variations we consider ε-parametrized families of transformations of *both* coordinates (parametrization) *and* fields such as

$$\left(X^{\mu}, \phi^{\alpha}\right) \to \left(\overline{X}^{\mu}, \bar{\phi}^{\alpha}\right), \tag{A.3}$$

with

$$\overline{X}^{\mu} = \kappa^{\mu}(\mathbf{X}, \varepsilon), \qquad \bar{\phi}^{\alpha}(X) = \Phi^{\alpha}\left(\phi^{\beta}(\mathbf{X}), \overline{\mathbf{X}}, \varepsilon\right), \tag{A.4}$$

where ε is an infinitesimal parameter such that for $\varepsilon = 0$ we have identically $\kappa^{\mu}(\mathbf{X}, 0) = X^{\mu}$, $\Phi^{\alpha}(\phi^{\beta}, \mathbf{X}, 0) = \phi^{\alpha}$. Here $\mathbf{X}$ is momentarily used to denote collectively the four space-time coordinates. It is assumed that the quantity L in (A.1) transforms as a *scalar* quantity, i.e.

$$L\left(\overline{\mathbf{X}}, \varepsilon\right) = \det\left(\partial\mathbf{X}/\partial\overline{\mathbf{X}}\right) L\left(\mathbf{X}\right). \tag{A.5}$$

We note that derivations with respect $\mathbf{X}$ and ε commute, and the same holds true of integration in $\mathbf{X}$ space and derivation with respect to ε. The variation of a field ϕ^{α} is then defined by

$$\delta\phi^{\alpha} := \partial\Phi^{\alpha}/\partial\varepsilon\big|_{\varepsilon=0}. \tag{A.6}$$

With vanishing fields at infinity in space and vanishing variations at the ends of the time interval, limiting ourselves to a first-order gradient theory, and applying an ε-parametrization to (A.1) we immediately have:

$$\delta A = \int d^{4}X \left(\sum_{\alpha}\left\{\frac{\partial L}{\partial\phi^{\alpha}}\delta\phi^{\alpha} + \frac{\partial L}{\partial(\partial_{\mu}\phi^{\alpha})}\delta(\partial_{\mu}\phi^{\alpha})\right\} + \frac{\partial L}{\partial X^{\mu}}\delta X^{\mu}\right). \tag{A.7}$$

In order that δA vanish for all admissible $\delta\phi^{\alpha}(\mathbf{X})$, and any α, with $\mathbf{X}$ *fixed*, a classical computation yields the following *Euler–Lagrange equations*:

$$E_{\alpha} \equiv \frac{\delta L}{\delta\phi^{\alpha}} = \frac{\partial L}{\partial\phi^{\alpha}} - \partial_{\mu}\frac{\partial L}{\partial(\partial_{\mu}\phi^{\alpha})} = 0, \tag{A.8}$$

for each $\alpha = 1, 2, \ldots, N$, at any space-time event $\mathbf{X}$. Equation (A.8) is a *strict* conservation law when L does not depend explicitly on ϕ^{α}. We note that the result (A.8) is left unchanged if we make the substitution

$$L \to L + \partial_{\mu}\Omega^{\mu}(\mathbf{X}),$$

i.e., if we add to L a four-divergence contribution in which Ω depends at most on the spacetime coordinates, and *not* on the fields.

The second group of equations, called *conservation equations*, that can derive from (A.1) by variation, results from a simultaneous transformation of both the coordinates X^μ *and* the fields. We shall not repeat all the details of the derivation of the resulting theorem known as *Noether's theorem* (Noether (1918); Soper (1976); Nelson (1979); see also Maugin (1993, pp. 97–118), or Maugin (2011b, pp. 78–103)). This celebrated theorem states that: *to any symmetry of the system there corresponds the conservation (or lack of strict conservation) of a current.* For L given in (A.1), such a current generally reads

$$J^\mu = L \frac{\partial \overline{X}^\mu}{\partial \varepsilon} + \sum_\alpha \frac{\partial L}{\partial(\partial_\mu \phi^\alpha)} \left[\frac{\partial \bar{\phi}^\alpha}{\partial \varepsilon} - \partial_\nu \phi^\alpha \frac{\partial \overline{X}^\nu}{\partial \varepsilon} \right], \qquad (A.9)$$

where ε must be taken equal to zero. In spite of the notation J^μ is not always simply a four-vector; it all depends on the group of transformations considered.

We just note the following steps. By invariance of the original action we mean that

$$\overline{A} = A(\varepsilon) = A(0), \quad \forall \varepsilon, \forall \phi^\alpha. \qquad (A.10)$$

Therefore, we need to find out the conditions in which $\partial A(\varepsilon)/\partial \varepsilon = 0$. Since

$$\delta A(\varepsilon) = \frac{\partial A}{\partial \varepsilon} \delta \varepsilon = \int d^4 X \, \delta L, \qquad \delta L \equiv \left. \frac{\partial L}{\partial \varepsilon} \right|_{\mathbf{X} \text{ fixed}} \delta \varepsilon, \qquad (A.11)$$

a *sufficient* condition for δA to vanish is that δL be the four-dimensional divergence of a certain quantity, noted δB^μ, in such a way that

$$\delta L = \partial_\mu (\delta B^\mu). \qquad (A.12)$$

But we can also compute δL directly. Using the second of (A.11) we have (omitting the α's)

$$\delta L = \frac{\partial L}{\partial \phi} \delta \phi + \partial_\mu \left(\frac{\partial L}{\partial(\partial_\mu \phi)} \delta \phi \right), \qquad (A.13)$$

up to irrelevant terms. Accounting for (A.8), we are entitled to write

$$(\partial_\mu J^\mu) \delta \varepsilon = -\frac{\delta L}{\delta \phi^\alpha} \delta \phi^\alpha, \qquad (A.14)$$

where the so-called *current* J^μ is such that

$$J^\mu \delta \varepsilon = -\delta B^\mu + \frac{\partial L}{\partial(\partial_\mu \phi^\alpha)} \delta \phi^\alpha. \qquad (A.15)$$

If the Euler–Lagrange equations (A.8) hold good, then (A.14) tells that the current is conserved. This is the contents of *Noether's theorem*. However, its remains to find the explicit expression of the current for each of the transformations involved in the set (A.3)–(A.4). This is in general obtained by evaluating

$$\delta\phi^\alpha(\mathbf{X}) = \left.\frac{\partial\bar\phi^\alpha}{\partial\varepsilon}\right|_{\overline{\mathbf{X}}\text{ fixed}}\delta\varepsilon = \left.\left(\frac{\partial\bar\phi^\alpha}{\partial\phi^\beta}\frac{\partial\phi^\beta}{\partial X^\mu}\frac{\partial X^\mu}{\partial\varepsilon}\right|_{\overline{\mathbf{X}}\text{ fixed}} + \frac{\partial\phi^\alpha}{\partial\varepsilon}\right)\delta\varepsilon. \quad \text{(A.16)}$$

For $\varepsilon = 0$ we note that $\partial X^\mu/\partial\varepsilon|_{\overline{\mathbf{X}}\text{ fixed}} = -\partial\overline{X}^\mu(\mathbf{X},\varepsilon)/\partial\varepsilon$. Accounting for this in (A.16) and (A.15) we obtain the announced general expression for the current:

$$J^\mu = L\frac{\partial\overline{X}^\mu}{\partial\varepsilon} + \frac{\partial L}{\partial(\partial_\mu\phi^\alpha)}\left(\frac{\partial\bar\phi^\alpha}{\partial\varepsilon} - \partial_\nu\phi^\alpha\frac{\partial\overline{X}^\nu}{\partial\varepsilon}\right). \quad \text{(A.17)}$$

Example: Space-time translations

In this case the transformations (A.3)–(A.4) reduce to a pure space-time transformation of the type

$$X^\mu \to \overline{X}^\mu(\mathbf{X},\varepsilon) = X^\mu + \varepsilon\,\delta^\mu_{.\lambda}, \quad \text{(A.18)}$$

where δ^μ_λ is the Kronecker symbol, i.e., it equals one only when μ takes the value granted to λ. There is no transformation of the fields themselves. On account of (A.17), we obtain the following current in components:

$$J^\mu \to J^\mu_{.\lambda} = T^\mu_{.\lambda} = L\,\delta^\mu_{.\lambda} - \sum_\alpha \partial_\lambda\phi^\alpha\frac{\partial L}{\partial(\partial_\mu\phi^\alpha)}. \quad \text{(A.19)}$$

This is a mixed second-order space-time tensor called the *energy-momentum tensor*. This denomination is clearly understood when we separate space and time components. If L does not depend explicitly on $\mathbf{X}$, then this tensor satisfies the following (four-dimensional) *strict conservation law*:

$$\partial_\mu T^\mu_{.\lambda} = 0, \quad \lambda = 1, 2, 3, 4. \quad \text{(A.20)}$$

Had we considered an explicit dependence of L on the space-time coordinates X^μ, instead of the strict conservation law (A.20) we would have a source term

$$f_\lambda = -\left.\frac{\partial L}{\partial X^\lambda}\right|_{\text{expl}}, \quad \text{(A.21)}$$

in the right-hand side.

With the *Cartesian notation* introduced in (A.2) we see that the independence of L on t, $\lambda = 4$, yields a *scalar* conservation law in the explicit form

$$\frac{\partial H}{\partial t}\bigg|_X - \nabla_R \cdot \mathbf{Q} = 0, \tag{A.22}$$

while the independence of L on the spatial part of $\mathbf{X}$, $\{X^K = 1, 2, 3\}$, yields the (covariant) *material balance law*:

$$\frac{\partial \mathbf{P}}{\partial t}\bigg|_X - div_R\, \mathbf{b} = \mathbf{0}, \tag{A.23}$$

where we defined the following canonical quantities:

- *energy (Hamiltonian density)*:

$$H := \sum_\alpha \dot{\phi}^\alpha \frac{\partial L}{\partial \dot{\phi}^\alpha} - L, \qquad \dot{\phi}^\alpha \equiv \frac{\partial \phi^\alpha}{\partial t}; \tag{A.24}$$

- *energy flux vector*:

$$\mathbf{Q} = \left\{ Q^K := -\sum_\alpha \dot{\phi}^\alpha \frac{\partial L}{\partial(\partial_K \phi^\alpha)} \right\}; \tag{A.25}$$

- *canonical (here material) momentum*:

$$\mathbf{P} = \left\{ P_K := -\sum_\alpha \frac{\partial \phi^\alpha}{\partial X^K} \frac{\partial L}{\partial(\partial \phi^\alpha / \partial t)} \right\}; \tag{A.26}$$

- *canonical stress tensor*:

$$\mathbf{b} = \left\{ b^K_{.L} := -\left(L\, \delta^K_L - \sum_\alpha \frac{\partial \phi^\alpha}{\partial X^L} \frac{\partial L}{\partial(\partial \phi^\alpha / \partial X^K)} \right) \right\}; \tag{A.27}$$

The "explicit" independence of L on t — no right-hand side in (A.22) – stands for the *conservation of energy*. The "explicit" independence of L on X^K signifies that the material body is *materially homogeneous*. But this is not a fundamental requirement of physics, so that, in general, (A.23) may contain a nonzero right-hand side denoted

$$\mathbf{f}^{\text{inh}} = \left\{ f^{\text{inh}}_L = \left(\frac{\partial L}{\partial X^L}\right)_{\text{expl}} \right\}, \tag{A.28}$$

so that (A.23) would be replaced by the inhomogeneous equation

$$\frac{\partial \mathbf{P}}{\partial t}\bigg|_X - div_R\, \mathbf{b} = \mathbf{f}^{\text{inh}}. \tag{A.29}$$

Remark A.6. Had we considered a Lagrangian depending explicitly on time, (A.22) would be replaced by

$$\frac{\partial H}{\partial t}\bigg|_X - \nabla_R \cdot \mathbf{Q} = h, \qquad h = -\frac{\partial L}{\partial t}\bigg|_{\text{expl}}, \qquad (A.30)$$

i.e., in a rheonomic system.

What must be principally gathered from the above is the essentially different nature of (i) the Euler–Lagrange equations (A.8) for which one such equation is written for *each field* — or in a more mechanical jargon, *each degree of freedom* — and of (ii) the *canonical equations of energy and momentum* — e.g., (A.22) and (A.23) — which pertain to the *whole* physical system and by necessity consider all fields simultaneously [note the summation over α in the definitions (A.24) through (A.27)]. In particular, in establishing (A.20) we have made use of the celebrated *Noether's identity* (A.14) here written as

$$(\delta_\mu J^\mu)\, \delta\varepsilon + \sum_\alpha (E_\alpha\, \delta\phi^\alpha) = 0, \qquad (A.31)$$

which emphasizes the above-made remark, but naturally implies (A.20) whenever all *field* equations (A.8) are satisfied simultaneously. It can also be written as

$$\delta_\mu J^\mu = 0 = \sum_\alpha E_\alpha \left(\frac{\partial \phi^\alpha}{\partial \varepsilon} - \frac{\partial \phi^\alpha}{\partial X^\beta}\frac{\partial X^\beta}{\partial \varepsilon} \right), \qquad (A.32)$$

with

$$J^\mu = T^\mu_{..\beta}\frac{\partial \overline{X}^\beta}{\partial \varepsilon} + \frac{\partial L}{\partial(\partial\phi^\alpha/\partial X^\mu)}\frac{\partial \phi^\alpha}{\partial \varepsilon},$$
$$T^\mu_{..\beta} = L\,\delta^\mu_\beta - \frac{\partial \phi^\alpha}{\partial X^\beta}\frac{\partial L}{\partial(\partial\phi^\alpha/\partial X^\mu)}. \qquad (A.33)$$

This is entirely equivalent to the foregoing formulation.

The above-given formulation is called *canonical* because it does not depend on the precise physical meaning of the fields ϕ^α. In standard elasticity the ϕ^α's are the three components of the placement or the three components of the displacement. In pure electromagnetism the ϕ^α's are made of the components of the electromagnetic potential. In generalized continua the ϕ^α's comprise the classical motion and that of the internal degrees of freedom, *etc.*

Appendix B

Appendix to Chapter 4: Justification of (4.33)–(4.34) by a two-timing method

Here we essentially follow the work of Whitham (1970). In this method of perturbation it is explicitly assumed that the time changes of the dependent variables occur at two different scales, fast oscillations being associated with a wave train and slow variations with the typical wave parameters (frequency, wave number, amplitude). Call T the slow time scale so that t, being the fast time, $\varepsilon = T/t$ is an infinitesimally small parameter. Corresponding length scales X and x are introduced (here we consider 1D problems for the sake of simplicity in the formalism). A typical dependent function, such as an elastic displacement, for such dynamical problems $u(x,t)$ is represented in the form

$$u(x,t) = U(X,T,\varphi), \tag{B.1}$$

where

$$X = \varepsilon\, x, \qquad T = \varepsilon\, t, \qquad \varphi = \varepsilon^{-1}\, \Phi(X,T). \tag{B.2}$$

Here φ is a *phase* with frequency ω and wave number k introduced by

$$\omega(X,T) = -\frac{\partial\varphi}{\partial t} = -\varphi_t = -\frac{\partial\Phi}{\partial T} = -\Phi_T,$$

$$k(X,T) = \frac{\partial\varphi}{\partial x} = \frac{\partial\Phi}{\partial X} = \Phi_X. \tag{B.3}$$

This shows that the scaling has been chosen in such a way that

$$\frac{\partial\omega}{\partial t} = \varepsilon\,\frac{\partial\omega}{\partial T}, \quad \frac{\partial\omega}{\partial x} = \varepsilon\,\frac{\partial\omega}{\partial X}, \quad \frac{\partial k}{\partial x} = \varepsilon\,\frac{\partial k}{\partial X}, \quad \frac{\partial k}{\partial t} = \varepsilon\,\frac{\partial k}{\partial T}. \tag{B.4}$$

From this we deduce the useful partial derivatives

$$u_t = -\omega\, U_\varphi + \varepsilon\, U_T, \qquad u_x = k\, U_\varphi + \varepsilon\, U_X. \tag{B.5}$$

In summary, the dependent function U is no longer a strictly periodic solution and with quantities X, T, ω, k, U all of order one in this perturbation scheme, the scaling is such that both ω and k vary slowly, and the field u has slow variation in addition to its oscillation with the phase φ. Now we apply these considerations to a Hamiltonian–Lagrangian variational principle of the type

$$\delta \int_x \int_t L(u_t, u_x, u)\, dx\, dt = 0, \tag{B.6}$$

with vanishing conditions at infinities. Using the short hand notation of Whitham (1970):

$$L_1 = \frac{\partial L}{\partial u_t}, \qquad L_2 = \frac{\partial L}{\partial u_x}, \qquad L_3 = \frac{\partial L}{\partial u}, \tag{B.7}$$

the field equation (standard Euler–Lagrange equation) reads

$$\frac{\partial}{\partial t} L_1 + \frac{\partial}{\partial x} L_2 - L_3 = 0. \tag{B.8}$$

Implementing now the scaling as in (B.5), this transforms to

$$\left(-\omega \frac{\partial}{\partial \varphi} L_1 + k \frac{\partial}{\partial \varphi} L_2 - L_3 \right) - \varepsilon \left(\frac{\partial}{\partial T} L_1 - \frac{\partial}{\partial X} L_2 \right) = 0. \tag{B.9}$$

Here the L_α, $\alpha = 1, 2, 3$ are functions

$$L_\alpha = L_\alpha(-\omega U_\varphi + \varepsilon U_T, k U_\varphi + \varepsilon U_X, U). \tag{B.10}$$

Equation (B.9) is a second-order equation for the function $U(X, T, \varphi)$ in (B.1).

As rightly noted by Whitham, the *art* (more than a rigorous reasoning) of the two-timing technique is to solve equation (B.9) by treating X, t, φ as three *independent* variables, so that $U(X, T, \varepsilon^{-1}\Phi)$ be a solution of the original problem. This solution is looked for as an expansion in ε:

$$U(X, T\varphi) = \sum_{n=0}^{\infty} \varepsilon^n U_n(X, T\varphi). \tag{B.11}$$

Any flexibility in the procedure is used to eliminate secular terms in φ whose presence would hinder the uniform validity of this expansion.

From Equation (B.9) we obviously obtain at the lowest order of approximation (indexed zero)

$$\frac{\partial}{\partial \varphi}\left(-\omega L_1^{(0)} + k L_2^{(0)} \right) - L_3^{(0)} = 0, \tag{B.12}$$

with

$$L_\alpha^{(0)} = L_\alpha\left(-U \omega_{0\varphi}, k U_{0\varphi}, U_0 \right). \tag{B.13}$$

Equation (B.12) can be integrated providing a first integral

$$\left(-\omega L_1^{(0)} + k L_2^{(0)} \right) U_{0\varphi} - L^{(0)} = A(X,T), \qquad (B.14)$$

with integration "constant" A, still a function of X and T. The latter is none other than the Hamiltonian or energy density of zeroth order defined by the following Legendre transformation in terms of the phase φ:

$$A = H\left(U, \Pi = \frac{\partial L}{\partial U_\varphi} \right) = U_\varphi \Pi - L^{(0)} = U_\varphi \frac{\partial L^{(0)}}{\partial U_\varphi} - L^{(0)}, \qquad (B.15)$$

and

$$\frac{\partial L^{(0)}}{\partial U_\varphi} = \frac{\partial L^{(0)}}{\partial u_t} \frac{\partial u_t}{\partial U_\varphi} + \frac{\partial L^{(0)}}{\partial u_x} \frac{\partial u_x}{\partial U_\varphi} = -\omega L_1^{(0)} + k L_2^{(0)}, \qquad (B.16)$$

according to (B.5) at zeroth order. But here A remains a function of the slow variables X and T. This means that the resulting solution at the lowest order of approximation looks like a periodic solution but still with the slow variations imposed by the present variables X and T. Note that the knowledge of A is equivalent to that of the amplitude of the wave.

One could proceed with the next order of approximation in order to obtain equations for ω, k and A. In particular, an equation can be deduced for the solution U_1 while the equations for ω, k and A are obtained from the conditions assuring us of the elimination of a secular term proportional to φ in that solution U_1. But instead of performing this typical procedure, we here focus attention on the expression of the variational principle (B.6). If with (B.5) we define

$$\widetilde{L} = \frac{1}{2\pi} \int_0^{2\pi} L\left(-\omega U_\varphi + \varepsilon U_T, k U_\varphi + \varepsilon U_X, U \right) d\varphi, \qquad (B.17)$$

then at lowest order this yields

$$\widetilde{L} = \frac{1}{2\pi} \int_0^{2\pi} L\left(-\omega U_{0\varphi}, k U_{0\varphi}, U_0 \right) d\varphi, \qquad (B.18)$$

where U_0 is the periodic solution extended so as to allow for a dependence of ω, k and A on X and T as in Equation (B.14), i.e., $\widetilde{L} = \widetilde{L}(\omega, k, A)$. Then, to lowest order, the principle (B.6) reads

$$\delta \iint \widetilde{L}(\omega, k, A) \, dX \, dT = 0, \qquad (B.19)$$

with (*cf.* (B.3))

$$\omega = -\Phi_T, \qquad k = \Phi_X \qquad (B.20)$$

This is the essence of the "averaged Lagrangian" method invented by Whitham (more details in Whitham (1970; 1974)). With appropriate generalization this method is well adapted to the treatment of nonlinear waves such as bright solitons (localized envelope signals with slowly modulated amplitude; *cf.* Benney and Newell (1967); Newell (1985); Maugin and Hadouaj (1991)).

Bibliography

Ablowitz, M. J. and Segur, H. (1981). *Solitons and the inverse scattering transform* (SIAM, Philadelphia).

Achenbach, J. D. (1973). *Wave propagation in elastic solids* (North-Holland, Amsterdam).

Aifantis, E. C. (1992). On the role of gradients in the localization of deformation and fracture, *Int. J. Egng. Sci.*, 30, pp. 1279–1299.

Andreev, N. N. (1995). On some second-order quantities in acoustics, *Sov. Phys. Acoust.*, 1(1), pp. 3–11, (reprint in *Sov. Phys. Acoust.*, 41(5), 684–689; original in Russian).

Andrejew, N. (1940). Uber die Energieausdrucke in der Akustik. *Journ. of Physics USSR*, 2, pp. 305–312.

Andrew, D. G. and McIntyre, M. E. (1978). On wave-action and its relatives, *J. Fluid Mech.*, 89, pp. 647–664.

Askes, H. and Aifantis, E. C. (2011). Gradient elasticity in statics and dynamics: an overview of formulations, length scale identification procedures, finite element implementations and new results, *Int. J. Solids Structures*, 48/13, pp. 1962–1990.

Auld, B. A. (1973). *Acoustic fields and waves in solids*, Two volumes (Wiley-Interscience, New York).

Barré de Saint-Venant A. (1869). Note sur les valeurs que prennent les pressions dans un solide élastique isotrope lorsque l'on tient compte des dérivées d'ordre supérieur des déplacements très petits que leurs points ont prouvés, *C. R. Acad. Sci. Paris*, 68, pp. 569–571.

Bathnagar, P. L. (1979). *Nonlinear waves in one-dimensional dispersive systems* (Oxford University Press, UK).

Benney, D. J. and Newell A. C. (1967). The propagation of nonlinear wave envelopes, *J. Math. and Phys.* (thereafter *Studies in Applied Mathematics*), 46, pp. 133–139.

Berezovski, A., Engelbrecht, J. and Maugin, G. A. (2008). *Numerical simulation of waves and fronts in inhomogeneous solids* (World Scientific, Singapore).

Bertoni, H. L. and Tamir T. (1973). Unified theory of Rayleigh-angle phenomena for acoustic beams at liquid-solid interfaces, *Appl. Phys. A. Materials*

Science & Processing, 2/4, pp. 157–172.

Bland, D. R. (1969). *Nonlinear dynamic elasticity* (Blaisdell, Waltham, Mass. USA).

Blekhman, I. I. and Lurie, K. A. (2000). On dynamic materials, *Doklady Akademii Nauk.*, 371, pp. 182–185 (in Russian).

Bleustein, J. L. (1968). A new surface wave in piezoelectric materials, *Appl. Phys. Lett.*, 13, pp. 412–414.

Blount, E. I. (1971). Stress-energy-momentum tensors in electromagnetic theory, *Bull. Telephone Laboratories*, Techn. Memo. No. 38139–9, Murray Hill, N.J.

Born, M. and Huang, K. (1954). *Dynamical theory of crystal lattices* (Oxford University Press, Oxford, U.K).

Brekhovskikh, L. M. (1960). *Waves in layered media* (Academic Press, New York/London).

Brenig, W. (1955). Bessitzen Schallwellen einen Impulz, *Zeit. Phys.*, 143, pp. 168–172.

Brillouin, L. (1921). La théorie des solides et les quanta, *Doctoral Thesis in Physics*, Faculté des Sciences de Paris, Gauthier-Villars, Paris [Jury: Marie Curie, Jean Perrin and Paul Langevin].

Brillouin, L. (1925). Sur les tensions de radiation, *Ann. Phys.*, 4, pp. 528–586.

Brillouin, L. (1960). *Wave propagation and group velocity* (Academic Press, New York).

Broglie, L. de (1924). Recherches sur la théorie des quanta, *D. Sc. in Physics*, Paris.

Calogero, F. and Degasperis, A. (1982). *Spectral transform and solitons: tools to solve and investigate nonlinear evolution equations*, Vol. I (North-Holland, Amsterdam).

Caloi, P. (1950). Comportement des ondes de Rayleigh dans un milieu firmo-élastique indéfini, *Publ. Bureau Central Seismol. Internat., Sér. A. Travaux scientifiques*, 17, pp. 89–108.

Carcione, J. M. (2007). Rayleigh waves in isotropic viscoelastic media, *Geoph. J. Intern.*, 108, pp. 453–464.

Chadwick, P. (1976). *Continuum mechanics* (George Allen and Unwin, London).

Christov, C. I. and Maugin, G. A. (1993). Long-time evolution of acoustic signals in nonlinear crystals, in: *Advances in nonlinear acoustics*, Ed. Hobaek, H. (World Scientific, Singapore), pp. 457–462.

Christov, C. I., Maugin, G. A., Velarde, M. G. (1996). Well-posed Boussinesq paradigm with purely spatial higher-order derivatives, *Phys. Rev. E*, 54/4, pp. 3621–3638.

Christov, C. I., Maugin, G. A. and Porubov, A. V. (2007). On Boussinesq's Paradigm in nonlinear wave propagation, Invited Contribution to Special Issue on J.V. Boussinesq, Ed. P. A. Bois, *C. R. Mécanique (Acad. Sci. Paris)*, 335, 9/10, pp. 21–535.

Ciarlet, P. G. (1988). *Mathematical elasticity*, Vol. I: Three-dimensional elasticity (North-Holland, Amsterdam).

Coquin, G. A. and Tiersten, H. F. (1967). Rayleigh waves in linear elastic dielectrics, *J. Acoust. Soc. Amer.*, 41, pp. 921–939.

Cosserat, E. and Cosserat, F. (1909). *Théorie des corps déformables* (Hermann, Paris).

Curie, P. K. and O'Leary, P. M. (1978). Viscoelatic Rayleigh waves II, *Q. Appl. Math.*, 35, pp. 445–454.

Curie, P. K., Hayes, M. A. and O'Leary, P. M. (1977). Viscoelastic Rayleigh waves, *Q. Appl. Math.*, 35, pp. 35–53.

Danilov, S. D. and Mironov, M. A. (2000). Mean force on a small sphere in a sound field in a viscous fluid, *J. Acoust. Soc. Am.*, 107/1, pp. 143–153.

Dauxois, T. and Peyrard M. (2006). *Physics of solitons* (Cambridge University Press, UK).

Dieulesaint, E. and Royer, D. (2000). *Elastic waves in solids* (translated from the French). (J. Wiley, New York).

Drazin, P. G. and Johnson, R. S. (1989). *Solitons: an introduction* (Cambridge University Press, UK).

Duhem, P. (1893). Le potentiel thermodynamique et la pression hydrostatique, *Ann. Ecole Normale* (3), 10, pp. 187–230.

Einstein, A. (1905). Über einen die Erzeugung und Verwandlung des Lichtes betreffenden heuristischen Gesichtspunkt, *Ann. Der Physik*, 17, pp. 132–148.

Epstein, M. and Maugin, G. A. (2000). Thermomechanics of volumetric growth in uniform bodies *Int. J. Plasticity*, 16, pp. 951–978.

Ericksen, J. L. (1977). Special topics in elastostatics, in: *Advances in applied mechanics*, Ed. C.-S. Yih, Vol. 17, (Academic Press, New York) pp. 189–244.

Ericksen, J. L. (1991). *Introduction to the thermodynamics of solids* (Chapman & Hall, London).

Eringen, A. C. (1967). *Mechanics of continua* (J. Wiley and Sons, New York).

Eringen, A. C. (1968). Theory of micropolar elasticity, in: H. Liebowitz (Ed), *Fracture: a treatise* (Academic Press, New York). Vol. II, pp. 621–729.

Eringen, A. C. (Editor, 1971-1976). *Continuum physics*, A series of four volumes (Academic Press, New York).

Eringen, A. C. (1972). Linear theory of nonlocal elasticity and dispersion of plane waves, *Int. J. Engng. Sci.*, 10, pp. 425–435.

Eringen, A. C. (1973). On Rayleigh surface waves with small wave lengths, *Lett. Appl. Engng. Sci.*, 1, pp. 11–17.

Eringen, A. C. (1974). Nonlocal elasticity and waves, in: *Continuum mechanics aspect of geodynamics and rock fracture mechanics*, Ed. Thoft-Christensen, P. (Reidel Publ. Co., Dordrecht, The Netherlands), pp. 81–105.

Eringen, A. C. (1980). *Mechanics of continua*, Revised and augmented edition (Krieger, Melbourne, Florida).

Eringen, A. C. (1987). Nonlocal dispersion of lattice dynamics and applications, in: *Constitutive models of deformation*, Eds. Chandra, J. and Srivastare, R. P. (SIAM, Philadelphia), pp. 58–80.

Eringen, A. C. (1999). *Microcontinuum field theories I: foundations and solids* (Springer, New York).

Eringen, A. C. (2002). *Nonlocal continuum field theories* (Springer, New York).

Eringen, A. C. and Maugin, G. A. (1990a). *Electrodynamics of continua*, Vol. I (Springer-Verlag, New York) [Soft cover reprint edition of the original edition, 2012].

Eringen, A. C. and Maugin, G. A. (1990b). *Electrodynamics of continua*, Two volumes (Springer-Verlag, New York).

Fokas, A. S. (1979). Generalized symmetries and constants of motion of evolution equations, *Lett. Math. Physics*, 3, pp. 467–473.

Frenkel, J. and Kontorova, T. (1938). On the theory of plastic deformation and twinning. *Physik Sowjet Union*, 123: pp. 1–15.

Georgiadis, H., Vardoulakis, I. and Velgaki, E. G. (2004). Dispersive Rayleigh-wave propagation in microstructured solids characterized by dipolar gradient elasticity, *J. Elasticity*, 74, pp. 17–45.

Germain, P. (1973). La méthode des puissances virtuelles en mécanique des milieux continus, Première partie: théorie du second gradient, *J. Mécanique (Paris)*, 12, pp. 235–274.

Ginzburg, V. L. and Ugarov, V. A. (1976). Remarks on forces and energy-momentum tensor in macroscopic electronics, *Sov. Phys. Usp.*, 19/1, pp. 94–101.

Ginzburg, V. L. and Tsytovich, V. N. (1979, English translation). Several problems of the theory of transition radiation and transition scattering, *Phys. Rep.*, pp. 49, 1–89 [Original Russian in: *Usp. Fiz. Nauk.*, 126, 553 (1978)].

Gubanov, A. (1945). Rayleigh waves on a boundary between a solid and a liquid, *J. Eksp. Teoret. Fiz. (USSR)*, 15, p. 497 (in Russian).

Gulyaev, Yu. V. (1969). Electroacoustic surface waves in solids, *Sov. Phys. JETP Lett.*, 9, pp. 35–38 (in Russian, 1968).

Gurevich, V. L. and Thellung, A. (1990). Quasi-momentum in the theory of elasticity and its conservation, *Phys. Rev.*, B42, pp. 7345–7349.

Gurevich, V. L. and Thellung, A. (1992). On the quasimomentum of light and matter and its conservation, *Physica*, A188, pp. 654–674.

Gurtin, M. E. and Murdoch, A. I. (1975) A continuum theory of elastic-material surfaces, *Arch. Rat. Mech. Anal.*, 57(4), pp. 291–323.

Hadouaj, H. and Maugin, G. A. (1992). Surface solitons on elastic structures: analysis and numerics, *Wave Motion*, 16, pp. 115–123.

Hadouaj, H., Malomed, B. A. and Maugin, G. A. (1991a). Dynamics of a soliton in the generalized Zakharov's System, *Phys. Rev.*, A44, pp. 3925–3931.

Hadouaj, H., Malomed, B. A. and Maugin, G. A. (1991b). Soliton-soliton collisions in the generalized Zakharov's system, *Phys. Rev.*, A44, pp. 3932–3940.

Hayes, W. D. (1970a). Conservation of action and modal wave action, *Proc. Roy. Soc. Lond.*, A320, pp. 187–206.

Hayes, W. D. (1970b). Kinematic wave theory, *Proc. Roy. Soc. Lond.*, A320, pp. 209–226.

Hayes, W. D. (1973). Group velocity and nonlinear dispersive wave propagation, *Proc. Roy. Soc. Lond.*, A332, pp. 199–221.

Hayes, W. D. (1974a). Introduction to wave propagation, in: *Nonlinear waves*, Eds. Leibvich, S. and Seebass, A. R. (Cornell University Press, NY) pp.

1–43.

Hayes, W. D. (1974b). Conservation of wave action, Chapter 6 in: *Nonlinear waves*, Eds. Leibovich, S. and Seebass, A. R. (Cornell University Press, Ithaca). pp. 170–185.

Henrich, V. E. and Cox, P. A. (1994). *The surface science of metal oxides* (Cambridge University Press, UK).

Holland, P. R. (1993). *The quantum theory of motion* (Cambridge University Press, U.K).

Infeld, I. and Rowlands, G. (1990). *Nonlinear waves, solitons and chaos* (Cambridge University Press, UK).

Jammer, M. (1974). *The philosophy of quantum mechanics* (J. Wiley-Science, New York).

Jones, J. P. and Whittier J. S. (1967). Waves at a flexibly bonded interface, *J. Appl. Mech.* (ASME), 34/4, pp. 905–908.

Jones, R. V. and Leslie, B. (1978). The measurement of optical radiation pressure in dispersive media, *Proc. Roy. Soc. Lond.*, A380, pp. 347–364.

Jones, R. V. and Richards, J. C. S. (1954). The pressure of radiation in a refracting medium, *Proc. Roy. Soc. Lond.*, A221, pp. 480–498.

Kalpakides, K. and Agiasofitou, E. K. (2002). On material equations in second-gradient electroelasticity, *J. Elasticity*, 67, pp. 205–227.

Kalyanasundaram, N. (1984). Nonlinear propagation characteristics of Bleustein–Gulyaev waves, *J. Sound and Vibrations*, 96, pp. 411–420.

Kittel, C. (2005). *Introduction to solid state physics* (Wiley International Edition, New York).

Kivshar, Yu. S. and Malomed, B. A. (1989). Dynamics of solitons in nearly integrable systems, *Rev. Mod. Phys.*, 61, pp. 763–915.

Knops, R. J., Trimarco, C. and Williams, H. T. (2003). Uniqueness and complementary energy in nonlinear elastostatics, *Meccanica*, 38, pp. 519–534.

Kosevich, A. M. (1999). *The crystal lattice (phonons, solitons, dislocations)* (Wiley-VCH, Berlin).

Kruskal, M. D. (1974). The Korteweg–de Vries equation and related evolution equations, in: *Nonlinear wave motion*, Ed. Newell, A. C., Lectures in Applied Mathematics, Vol. 15 (American Math. Soc., Providence, R.I), pp. 61–83.

Kruskal, M. D. and Zabusky, N. (1966). Exact invariants for a class of nonlinear wave equations, *J. Math. Phys.*, 7, pp. 1265–1267.

Kunin, I. A. (1982). *Elastic media with microstructure*, I & II (Springer-Verlag, Berlin), (translation supervised by E. Kröner from the 1975 Russian edition).

Lai, C. G. and Rix, G. L. (2002). Solution of the Rayleigh eigenproblem in viscoelastic media, *Bull. Seism. Soc. America*, 92, pp. 2297–2309.

Lanczos, C. (1962). *The variational principles of mechanics* (University of Toronto Press, Toronto).

Landau, L. D. and Lifshitz, E. M. (1965). *Mécanique* (translation from the Russian of the first volume of the Course on Theoretical Physics) (Editions MIR, Moscow).

Lazar, M. (2012). Non-singular dislocation loops in gradient elasticity, *Phys. Lett.*, A376, pp. 1757–1758.

Lazar, M. and Anastassiadis, C. (2007). Lie point symmetries, conservation laws and balance laws in linear gradient elastodynamics, *J. Elasticity*, 88, pp. 5–25.

Lazar, M. and Maugin, G. A. (2005). Nonsingular stress and strain fields of dislocations and disclinations in first strain gradient elasticity, *Int. J. Engng. Sci.*, 43, pp. 1157–1184.

Lazar, M. and Maugin, G. A. (2007). On microcontinuum field theories: the Eshelby stress tensor and incompatibility conditions, *Philos. Mag.*, 87/5, pp. 3853–3870.

Le Roux, J. (1911). Etude géométrique de la torsion et de la flexion, dans les déformations infinitésimales d'un milieu continu, *Ann. Ecole Norm. Sup.*, 28, pp. 523–579.

Le Roux, J. (1913). Recherches sur la géométrie des déformations finies, *Ann. Ecole Norm. Sup.*, 30, pp. 193–245.

LeBlond, P. H. and Mysak, L. A. (1978). *Waves in the ocean* (Elsevier, Amsterdam).

Lighthill, M. J. (1965). Contribution to the theory of waves in nonlinear dispersive systems, *J. Inst. Maths. Applics.*, 1, pp. 269–306.

Love, A. E. H. (1911). *Some problems of geodynamics* (Cambridge University Press, UK).

Ludwig, W. (1991). Dynamics at crystal surfaces, surface phonons, *Int. J. Engng. Sci.*, 29/3, pp. 345–361.

Lurie, K. A. (2007). *Introduction to the mathematical theory of dynamic materials* (Springer, New York).

Lurie, K. A. (2009). On homogenization of activated laminates in 1D-space and time, *Zeit. Angew. Math. Mech.*, 89/4, pp. 333–340.

Lurie, K. A., Onofrei, D. and Weekes, S. L. (2009). Mathematical analysis of the wave propagation through a rectangular material structure in space-time, *J. Math. Anal. Appl.*, 395/1, pp. 180–194.

Lurie, K. A. and Weekes, S. L. (2006). Wave propagation and energy exchange in a spatio-temporal material composite with rectangular microstructure, *J. Math. Anal.*, A314/1, pp. 286–310.

Mackenzie, K. V. (1960). Reflection of sound from coastal bottoms, *J. Acoust. Soc. America*, 32/2, pp. 221–231.

Marsden, J. E. and Hughes, T. R. (1975). *Mathematical theory of elasticity* (Academic Press, New York) [also as Dover reprint].

Maugin, G. A. (1988). *Continuum mechanics of electromagnetic solids* (North-Holland, Amsterdam; Volume 33 in the series: Applied Mathematics and Mechanics).

Maugin, G. A. (1993). *Material inhomogeneities in elasticity* (Chapman & Hall, London).

Maugin, G. A. (1998). On the structure of the theory of polar elasticity, *Phil. Trans. Roy. Soc. Lond.*, A356, pp. 1367–1395.

Maugin, G. A. (1999a). *The thermomechanics of nonlinear irreversible behaviors*

(World Scientific, Singapore).

Maugin, G. A. (1999b). *Nonlinear waves in elastic crystals* (Oxford University Press, U.K).

Maugin, G. A. (2006). On canonical equations of continuum thermomechanics, *Mech. Res. Commun.*, 33, pp. 705–710.

Maugin, G. A. (2007). Nonlinear kinematic wave mechanics of elastic solids, *Wave motion*, 44/6, pp. 472–481.

Maugin, G. A. (2008), On phase, action and canonical conservation laws in kinematic-wave theory, *Fizika Nizkikh Temperatur* (Ukraine; Issue dedicated to the late A.M. Kosevich), Vol. 34, No.7, pp. 721–724 (in Russian). (In English: *Low Temperature Physics*, 34, 7, 2008).

Maugin, G. A. (2009). On inhomogeneity, growth, ageing and the dynamics of materials, *J. Mech. Materials and Structure*, 4, pp. 731–741.

Maugin, G. A. (2011a). A historical perspective of generalized continuum mechanics, in: *Mechanics of generalized continua*, Eds. Altenbach, H., Maugin, G. A. and Erofeev, V. (Springer, Berlin), pp. 3–19.

Maugin, G. A. (2011b). *Configurational forces: thermomechanics, physics, mathematics, and numerics* (CRC/Chapam & Hall / Taylor and Francis, Boca Raton, FL).

Maugin, G. A. (2013). *Continuum mechanics through the twentieth century: a concise historical perspective* (Springer, Dordrecht).

Maugin, G. A. (2014). *Continuum mechanics through the eighteenth and nineteenth centuries: historical perspective from John Bernoulli (1727) to Ernst Hellinger (1914)* (Springer, Dordrecht).

Maugin, G. A. and Christov, C. I. (2002). Nonlinear duality between elastic waves and quasi-particles, in: *Selected topics in nonlinear wave mechanics*, Eds. Christov, C. I. and Guran, A. (Birkhäuser, Boston), pp. 117–160.

Maugin, G. A. and Epstein, M. (1991). The electroelastic energy-momentum tensor, *Proc. Roy. Soc. Lond.*, A433, pp. 299–312.

Maugin, G. A. and Hadouaj, H. (1991). Solitary surface transverse waves on an elastic substrate coated with a thin film, *Phys. Rev.*, B44, pp. 1266–1280.

Maugin, G. A., Hadouaj, H. and Malomed, B. A. (1992a). Nonlinear coupling between SH surface solitons and Rayleigh modes on elastic structures, *Phys. Rev.*, B45, pp. 9688–9694.

Maugin, G. A., Pouget, J., Drouot, R. and Collet, B. (1992b). *Nonlinear electromechanical couplings* J. Wiley, Chichester, UK, and New York, USA).

Maugin, G. A. and Rousseau, M. (2010a). On two neglected/ignored conservation laws in continuum mechanics: canonical momentum and action. Contribution to the 70th Anniversary Volume of K. Wilmanski: *Mechanics of continua with microstructure*, Ed. Albers, B. (Springer, Berlin), pp. 247–264.

Maugin, G. A. and Rousseau, M. (2010b). Bleustein–Gulyaev SAW and its associated quasi-particle, *Int. J. Engn. Sci.* (Special issue honouring K. Rajagopal), 48, pp. 1462–1469.

Maugin, G. A and Rousseau, M. (2012a). Grains of SAWs: associating quasi-particles to surface acoustic waves, *Int. J. Engng. Sci.*, (special issue in honor of V. L. Berdichevsky), 59, pp. 156–167.

Maugin, G. A. and Rousseau, M. (2012b). Wave-quasi-particle dualism in the transmission-reflection problem for elastic waves, *J. Theoret. Appl. Mech. (PL)* 50/3, pp. 797–805 (invited contribution in the jubilee issue of 2012).

Maugin, G. A. and Rousseau, M. (2013). Prolegomena to studies on dynamic materials and their space-time homogenization, *Discr. Contin. Dyna. Sys.*, Series S (DCDS-S), 6/6, pp. 1599–1607.

Maugin, G. A. and Trimarco, C. (1992). Pseudomomentum and material forces in nonlinear elasticity: variational formulations and application to brittle fracture, *Acta Mechanica*, 94, pp. 1–28.

Maynard, J. D. (2000). Phonons in crystals, Chapter 57 in: *Encyclopedia of Acoustics*, Ed. Crocker, M. J. (John Wiley, New York) Vol. II, pp. 657672.

McIntyre, M. E. (1981). On the "wave momentum" myth, *J. Fluid Mech.*, 106, pp. 331–347.

Miloserdova, I. V. and Potapov, A. I. (1983). Nonlinear standing waves in a rod of finite length. *Sov. Phys. Acoust.*, 29(4), pp. 515–520.

Mindlin, R. D. (1964). Microstructure in linear elasticity, *Arch. Rat. Mech. Anal.*, 16, pp. 51–78.

Mindlin, R. D. and Tiersten, H. F. (1962). Effects of couple stresses in linear elasticity, *Arch. Rat. Mech. Anal.*, 11, pp. 415–448.

Miura, R. M. (1974). The Korteweg–deVries equation: a model equation for nonlinear dispersive waves, in: *Nonlinear waves*, Eds. Leibovich, S. and Seebass, A. R. (Cornell University Press, Ithaca, NY), pp. 212–234.

Murdoch, I. A. (1976). The propagation of surface waves in bodies with material boundaries. *J. Mech. Phys. Solids*, 24, pp. 137–146.

Murnaghan, F. D. (1951). *Finite deformation of an elastic solid* (J. Wiley, New York).

Nadin, G. (2009). Travelling fronts in space-time periodic media, *J. Math. Pures Appl.*, 92, pp. 232–262.

Nelson, D. F. (1979). *Electric, optic and acoustic interactions in dielectrics* (Wiley-Interscience, New York).

Nelson, D. F. (1990). Resolution of the problem of Minkowski and Abraham, in: *Mechanical modelling of new electromagnetic materials*, Ed. Hsieh, R. K. T. (Elsevier, Amsterdam) pp. 171–177.

Nelson, D. F. (1991). Momentum, pseudomomentum and wave momentum: toward resolving the Minkowski-Abraham controversy, *Phys. Rev.*, A44, pp. 3905–3916.

Newell, A. C. (1985). *Solitons in Mathematics and Physics*, (SIAM, Philadelphia).

Noether, E. (1918). Invariante Variationsproblem, *Klg Ges. Wiss. Nach. Göttingen. Math. Phys.*, Kl.2, pp. 235–257 [English translation by M. Tavel: Invariant variation problems, *Transp. Theory Stat. Phys.*, 1, 186–207, 1971; French translation in the book: Y. Kosmann–Schwarzbach (2004), *Les théorèmes de Noether*, Editions de l'Ecole Polytechnique, Palaiseau, France].

Nowacki, W. (1986). *Theory of asymmetric elasticity* (Pergamon Press, Oxford, UK) (translated from the Polish).

Ogden, R. W. (1984). *Nonlinear elastic deformations* (Ellis Horwood, Chichester,

UK) [also as Dover reprint].

Ostrovsky, L. A. and Potapov, A. I. (1999). *Modulated waves* (Johns Hopkins University Press, Baltimore).

Peierls, R. (1976). The momentum of light in a refracting medium, *Proc. Roy. Soc. Lond.*, A347, pp. 475–491.

Peierls, R. (1979). *Surprises in theoretical physics* (Princeton Univ. Press, USA).

Peierls, R. (1985). Momentum and Pseudomomentum of light and sound, in: *Highlights of condensed-matter physics*, Ed. M. Tosi, Corso LXXXIX, pp. 237–255, Soc. Ital. Fisica, Bologna.

Peierls, R. (1991). *More surprises in theoretical physics* (Princeton Univ. Press, USA).

Piola, G. (1848). Intorno alle equazioni fondamentali del movimento di corpi quasivoglioni considerati secondo la naturale loro forma e costituva, *Mem. Math. Soc. Ital. Modena*, 24(1), pp. 1–186.

Porubov, A. V. and Maugin, G. A. (2009). Cubic nonlinearity and longitudinal surface solitary waves, *Int. J. Non-lin. Mech.*, 44, pp. 552–559.

Potapov, A. I. and Maugin, G. A. (2001a). Wave momentum and strain radiation in elastic bodies (in Russian), *Vestnik Nizheg. Univ. I.N. Lobachevskii*, Série Mechanics, 3, pp. 92–106.

Potapov, A. I. and Maugin, G. A. (2001b). Wave momentum in elastic solids, in: *Proc. Int. Conf. "Progress in nonlinear science"* (Nizhny-Novgorod, Russia, July 2001).

Potapov, A. I., Maugin, G. A. and Trimarco, C. (2005). Wave momentum and radiation stresses in elastic solids, *Mathematics and mechanics of solids*, 10, pp. 441–460.

Rayleigh, J. W. S. (1887). On waves propagated along the plane surface of an elastic solid, *Proc. Lond. Math. Soc.*, 17, pp. 4–11.

Rayleigh, Lord (1905). On the momentum and pressure of gaseous vibrations, and on the connexion with the virial theorem, *Phil. Mag.*, 10, pp. 364–374.

Rebi, C. and Soliani, G. (Eds, 1984). *Solitons and particles* (World Scientific, Singapore).

Romeo, M. (2001a). Rayleigh waves on a viscoelastic solid half-space, *J. Acoust. Soc. Amer.*, 110, pp. 59–67.

Romeo, M. (2001b). A solution for transient surface waves of the B-G type in dissipative piezoelectric crystal, *Zeit. Angew. Math. Mech. (ZAMP)*, 52/5, pp. 730–748.

Rosenau, P. and Hyman, J. M. (1993). Compactons: solitons with finite wavelength, *Phys. Rev. Lett.*, 70/5, pp. 564–567.

Rousseau, M. and Maugin, G. A. (2011a). Rayleigh surface waves and their canonically associated quasi-particles, *Proc. Royal Soc. Lond.*, A467, pp. 495–507.

Rousseau, M. and Maugin, G. A. (2011b). Bleustein–Gulyaev SAWs with low losses: approximate direct solution, *J. Electromag. Anal. Appl.*, 3, pp. 122–127.

Rousseau, M. and Maugin, G. A. (2012a). Influence of viscosity on the motion of quasi-particles associated with surface acoustic waves, *Int. J. Engng. Sci.*, 50(1), pp. 10–21.

Rousseau, M. and Maugin, G. A. (2012b). Quasi-particle aspects of Murdoch surface acoustic waves, *Quart. J. Mech. Applied Math.*, 65, pp. 333–345.

Rousseau, M. and Maugin, G. A. (2012c). Quasi-particles associated with Bleustein-Gulyaev SAWs: perturbations by elastic nonlinearities, *Int. J. Non-lin. Mech.*, 47, pp. 67–71.

Rousseau, M. and Maugin, G. A. (2013). Wave momentum in models of generalized continua. *Wave Motion*, 50(3), pp. 509–519.

Rousseau, M. and Maugin, G. A. (2014). The Love surface acoustic wave seen as a non-Newtonian quasi-particle (accepted for publication in 2015).

Rousseau, M., Maugin G. A. and Berezovski, M. (2011). Elements of study on dynamic materials, *Arch. Appl. Mech.*, 81, pp. 925–942.

Sanchez–Palencia, E. and Zaoui, A. (1987, Editors). *Homogenization techniques for composite media* Lecture Notes, Vol. 272 (Springer, Berlin).

Sanz-Serna, J. M. and Calvo, M. P. (1994). *Numerical Hamiltonian problems* (Chapman and Hall, London).

Scholte, J. G. (1947). On Rayleigh waves in visco-elastic media, *Physica (Utrecht)*, 13, pp. 245–250.

Smith, G. E. (2006). The vis-viva dispute: a controversy at the dawn of dynamics, *Physics Today*, 59/10, pp. 31–36

Soper, D. E. (1976). *Classical field theory* (Wiley, New York).

Spencer, A. J. M. (1976). *Continuum mechanics* (Longman, Harlow, UK).

Stepanyants, Yu. A. and Fabrikant, A. L. (1989). Wave propagation in shear hydrodynamic flows, *Sov. Phys. Usp.*, 32/9, 783–605.

Stepanyants, Yu. A. and Fabrikant, A. L. (1999). *Wave propagation in shear flows* (World Scientific, Singapore).

Stone, M. (2000). Phonons and forces: momentum versus pseudomomentum in moving fluids, arXiv:cond-mat/0012316 v1, http://arxiv.org/pdf/cond-mat/0012316.

Stoneley, R. (1924). Elastic waves at the surface of separation of two solids. *Proc. Roy. Soc. Lond.*, A106, pp. 416–428.

Suhubi, E. S. and Eringen A. C. (1964). Nonlinear theory of micro-elastic solids II, *Int. J. Engng. Sci.*, 2/4, pp. 389–404.

Tiersten, H. F. (1969). Elastic surface waves guided by thin films, *J. Appl. Phys.*, 40, pp. 770–789.

Toupin, R. A. (1964). Theory of elasticity with couple-stress, *Arch. Rat. Mech. Anal.*, 17, pp. 85–112.

Truesdell, C. A. and Noll, W. (1965). Nonlinear field theories of mechanics, in: *Handbuch der Physik*, Bd. III/3, Ed. Flügge, S. (Springer-Verlag, Berlin).

Truesdell, C. A. and Toupin, R. A. (1960). The classical field theories, in: *Handbuch der Physik*, Bd. III/1, Ed. Flügge, S. (Springer-Verlag, Berlin).

Tsai, Y. M. and Kolsky, H. (1968). Surface wave propagation for linear viscoelastic solids, *J. Mech. Math. Solids*, 16, pp. 99–109.

Vardoulakis, I. and Georgiadis, H. G. (1997). SH surface waves in a homogeneous gradient-elastic half-space with surface energy, *J. Elasticity*, 47, pp. 147–165.

Vesnitskii, A. I. and Metrikine, A. V. (1996). Transition radiation in mechanics.

Physics-Uspekhi, 39, pp. 983–1007 (English translation).

Vlasie–Belloncle, V. and Rousseau, M. (2006). Effect of surface free energy on the behaviour of surface and guided waves, *Ultrasonics*, 45, pp. 188–195.

Westervelt, P. L. (1957). Acoustic radiation pressure. *J. Acoust. Soc. Am.*, 29, pp. 26–29.

Whitham, G. B. (1965). A general approach to linear and nonlinear dispersive waves using a Lagrangian, *J. Fluid Mech.*, 22, pp. 273–283.

Whitham, G. B. (1970). Two-timing variational principles, and waves, *J. Fluid Mechanics*, 44, pp. 73–395.

Whitham, G. B. (1974). *Linear and nonlinear waves* (Interscience-J. Wiley, New York).

Wilson, E. B. (1914). Review of Love's book "Some problems in geodynamics", *Bull. Amer. Math. Soc.*, 20/6, pp. 432–434.

Zakharov, V. E. and Shabat, A. B. (1972). Exact theory of two-dimensional self-focusing and one-dimensional self-modulation in nonlinear media. *Soviet Physics J.E.T.P.*, 34, pp. 62–69.

Zarembo, L. K. and Krasil'nikov, V. A. (1966). *An introduction to nonlinear acoustics* (Nauka, Moscow; in Russian).

Index